ENDANGERED
ORCAS

ENDANGERED
ORCAS

THE STORY OF THE
SOUTHERN RESIDENTS

MONIKA WIELAND SHIELDS

Paperback ISBN: 978-1-7336934-0-0
eBook ISBN: 978-1-7336934-1-7

Library of Congress Control Number: 2019901681

All photographs by the author unless otherwise noted.

Front cover image by Monika Wieland Shields
Book design by *the*BookDesigners

First printing edition 2019

Orca Watcher
Friday Harbor, WA
publisher@orcawatcher.com

www.orcawatcher.com

For J2

CONTENTS

PROLOGUE

"THEY'RE GOING TO TRY TO FEED HER," the woman next to me said. "I hope she takes the fish."

I was sitting along the shoreline on the west side of San Juan Island, Washington, on an afternoon in August 2018, a summer that had been unlike any other. J50, a three-and-a-half-year-old female also known as Scarlet, was surfacing about a quarter-mile offshore, no other whales in sight. When Scarlet was born at the end of 2014, she was a beacon of hope for a population of killer whales that hadn't seen a successful birth in over two years. Her entry into the world was likely a dramatic one, as from day one she had deep scratches on her dorsal fin and back that led researchers to speculate she may have been partially pulled out of the womb by another whale. Those scars became her namesake. Known for her playful spunk and often wandering away from mom, Scarlet was the first of the "baby boom" whales, a spate of nine births in thirteen months, a number only seen once before in more than 40 years of study. Some hoped Scarlet's birth was the turning point for the endangered Southern Resident killer whales, but instead, things have changed for the worse.

Since the baby boom, another three years have passed with no successful births for the Southern Residents, and half of those born during that year have since died. When J-Pod returned to inland waters in June of 2018, Scarlet looked emaciated, showing a depression behind her blowhole known as "peanut head". Scientists gave her weeks to live, but two months later, while her condition has deteriorated, Scarlet was somehow still swimming. Near the end of July, another J-Pod whale, J35 Tahlequah, gave birth to a daughter that lived only a matter of minutes before dying, and the grieving mother spent the next 17 days carrying the body of her deceased neonate through the waters of Washington State and British Columbia in a vigil that gained international attention and turned the spotlight on this small population of whales that calls the Pacific Northwest home. As a result, Scarlet's story also made headlines, and NOAA Fisheries led an unprecedented collaborative effort to intervene in the fate of a wild whale.

First, a team of veterinarians and researchers assembled to collect breath and fecal samples from Scarlet to try and determine what was wrong with her. Though she was visibly malnourished, that didn't mean she was starving; a lack of food likely played into whatever was ailing her, but the rest of her family didn't appear emaciated, so the thought was she may have contracted a disease. While test results proved inconclusive, treatment plans went ahead, and several doses of antibiotics and de-wormers were administered via a dart gun. Now, on this August afternoon, a test run was underway to see if she would accept living Chinook salmon dropped off a chute from a boat in front of her, to evaluate both the success of supplemental feeding and explore the possibility of administering medications orally.

Scarlet, always small for her age, continued to breathe slowly, her back barely breaking the surface as she came up for air. Surrounding her were six enforcement and research vessels. No whale-watch boats were present as they voluntarily gave Scarlet and her family extra space during this ordeal, and no other whales were present as, mysteriously for these family-oriented orcas, they seemed content to leave their ailing member alone for hours or even days at a time. I watched as one of the response vessels slipped Chinook salmon off their stern and into the water. The intention was good, but the execution was questionable: there was no way to tell what happened to the fish after they entered the murky waters of Haro Strait, whether they were in fact caught and eaten or simply swam away.

Watching this unfold I had to stop to ask myself how we got here. Why were we talking about medicating or even capturing and rehabilitating a wild whale? Yes, we want Scarlet to live, but treating her illness is a band-aid solution to the real problems this unique population of killer whales is facing: first and foremost, declines of their preferred Chinook salmon prey along the West Coast. More importantly, where do we go from here? We have had three years with no births, more whales dying, and no changes to this trend in sight. If we want to continue to have orcas as our neighbors, is the path forward to individually treat one whale after another as they become sick and malnourished? Or do we need to have some bigger conversations about how we live in these shared waters, and what changes need to be made to ensure a healthy future not just for killer whales but for all creatures of the Salish Sea, ourselves included?

The circus surrounding J50 Scarlet continued for another four weeks as she continued to look worse and worse. Even when it was clear that she had finally succumbed to whatever was ailing her and died, there was a theatric show of searching for her by boat and helicopter for days before the intervention team was willing to admit that she was gone. With her death, the Southern Resident killer whale population now numbered just 74, the lowest in more than 30 years. Where, indeed, do we go from here?

INTRODUCTION

The Orca Watcher

I REMEMBER THE FIRST TIME I saw a superpod, a gathering of every orca in the Southern Resident population. My parents and I were aboard the *Bon Accord*, a wooden fishing boat out of Friday Harbor that in the 1980s had been one of the first whale-watching vessels in the region. For a full two hours we were shut down in northern Haro Strait as more than eighty whales from J-, K-, and L-Pods socialized in groups. Video camera in hand, I sat on the roof of the wheelhouse, not sure which way to look. Action surrounded us. Off in the distance a female breached, followed a moment later by her calf. On the port side a male lazily swam upside down, his pectoral fins in the air as he did one, two, three inverted tail slaps. Looking over my shoulder, I spotted a dozen dorsal fins at the surface facing every which way as the whales mingled and rolled around together. At one point three whales broke off from the rest and came right toward us. It was the K18 family—K18 Kiska and her likely daughter, K40 Raggedy, and son, K21 Cappuccino. The trio twice circled the boat, close enough that I could see their entire black-and-white bodies from nose to tail, their tail flukes barely moving as they swam effortlessly around us. Kiska surfaced directly below me, giving me a face full of spray from her exhalation. The captain, who had watched these whales for nearly twenty years, was at a loss for words: "You don't see this . . . you just don't see this."

I was sixteen and had found my life's passion.

My love of whales and dolphins spanned back to my early childhood, and killer whales (also known as orcas) were my standout favorite. In elementary school I drew orcas and wrote stories about people befriending killer whales. Visiting my grandparents in San Diego, I'd get my parents to take me to SeaWorld. I saw the Shamu show, but I wasn't impressed. Instead, I wanted to sit by the killer whale tanks. I wondered what they would do when they weren't being asked to perform. At age twelve, I was so devoted to observing them that my parents finally agreed to drop me off at the park on my own, leaving me to spend the whole day by myself at the killer whale enclosure.

Later that same summer, we took a trip to Alaska. It was the last vacation for our family of four, as my older brother had just graduated college and was soon getting married. It was a memorable trip for all of us, but the most profound moment for me occurred on a full-day boat tour in Glacier Bay National Park, where I first laid eyes on wild killer whales. As we pulled away from a calving glacier, I heard the announcement over the PA that a pod of orcas was approaching off our starboard side. I eagerly scanned the murky green waters, my heart getting a jolt when that initial tall dorsal fin split the surface of the water. The small group of whales traveled quickly in a tight group, and we stayed with them only a short while. On a second boat trip, we saw orcas again, and I took my first whale photos. I loved these whales, and I wanted more. Back at home in Portland, Oregon, I started researching to see if there was somewhere closer to home where I could see orcas. I learned about the Southern Resident killer whales and the San Juan Islands in Washington State, just a five-hour drive north from my hometown.

Two years after the Alaska trip, my parents took me up to Friday Harbor on San Juan Island for my birthday in October. Seeing whales is hit or miss that time of year, but we went out on the water anyway. I sat outside in the drizzle the entire time, scanning, but no luck. We returned the following June for a week, and the four whale-watch trips we took cemented the place of the Southern Residents in my heart. Every subsequent summer I found a way to spend more time on San Juan Island with the orcas. I met Bob Otis, who was doing shore-based whale behavioral studies at Lime Kiln Lighthouse, on the west side of San Juan Island, one of the best places in the world to see killer whales from shore. As a high school sophomore, I got myself a two-week internship with Bob. During this stint on the island I met Rich Osborne, who encouraged me to apply for an internship at The Whale Museum. For five years I interned at the museum, working mostly on an acoustic research project with the hydrophones out at Lime Kiln. I spent every summer day making the recordings that became the basis for my Reed College undergraduate senior thesis on killer whale acoustic communication. I filmed and photographed the whales, learning how to individually identify them. Eventually, I began working as a boat-based naturalist, primarily for Western Prince Whale and Wildlife Tours out of Friday Harbor.

After college, I moved to San Juan Island full-time, published a book of orca photographs, and started my blog, *Orca Watcher*, featuring descriptions of my whale sightings and photos. For six years I worked seasonally as a naturalist with Western Prince, then took a part-time job as a lab technician at a small manufacturing company on the island. This gave me flexible, year-round work that would help me pay the bills while I pursued my real passion: the whales. I was a charter member of the Salish Sea Association of Marine Naturalists, taught the acoustics segment of The Whale Museum's biannual Marine Naturalist Training Program, continued blogging and taking and sharing photos, and started writing more about the whales. I bought my own boat, and with a fellow Reed College graduate, I cofounded the nonprofit Orca Behavior Institute with a focus on conducting independent field-based behavioral and acoustic research.

Ever since I first saw these whales, they have been a central factor in my life.

The Southern Residents are the most watched and most studied population of killer whales in the world. There is a lot of information out there about them, from Facebook photos to sighting reports to scientific journal articles, but there has been little telling of their whole story in one place. In 1981, Erich Hoyt published the iconic *Orca: The Whale Called Killer*, in which he recounted how the 1970s transformed people's relationship to killer whales in the Pacific

The author photographing killer whales from shore at Lime Kiln Point State Park on San Juan Island. Photo by Jason Shields.

Northwest. My vision was to write in the same vein, interweaving stories of individuals and specific whales, with further scientific discoveries covering the nearly forty years since Hoyt's book. Initially intimidated by undertaking this task, but with encouragement from others, I began to collect my writing, based on my own experiences, with the idea of compiling a book. I realized that telling the complete story of the Southern Residents would involve interviewing others. As I did so, I gained an appreciation for my unique perspective on these whales. Given my various experiences, I have connections in the research community, in the whale-watch industry, with marine educators, and with shore-based whale-watchers. In a sometimes divided whale community—perhaps like any small dedicated group with its own set of internal politics—I realized that I had the broad connections necessary to tell a complete narrative with varied perspectives.

This book is divided into three parts. The first part provides an introduction to the complex species of killer whales in general and the Southern Residents in particular, covering the different types of killer whales found around the world, the life history and social system that make killer whales unique in the animal kingdom, the fascinating world of acoustic communication, and the controversial topic of personality, emotions, and culture in nonhumans. The second part characterizes our changing relationship with the Southern Residents over the past six decades, beginning with a discussion of the Salish Sea, where orcas and humans come together. We follow the story of a species once feared and despised, transformed by the popularity of being captured for entertainment into one extensively studied and observed in the wild. Exploring the people, passion, and politics that surround the extraordinary Southern Residents, I look at the positive attachment and negative behaviors that come hand in hand with watching these whales. Finally, the third part of the book tells the story of a whale population in trouble, describing the risk factors that led to the decline of the Southern Residents, how current research has informed conservation issues, and what actions may lay the groundwork for recovery of this endangered population.

PART 1

UNDERSTANDING KILLER WHALES

CHAPTER 1

A Complex Species
Different Kinds of Killer Whales

KNOWN AS ONE OF THE world's best places for seeing killer whales from land, Lime Kiln Point State Park sits on the western coast of San Juan Island in Washington State. Off the quarter-mile rocky shoreline that stretches from the Lime Kiln Lighthouse at the park's north end to an observation wall at the south end, the land drops off steeply, making the water deep enough for whales to swim within a few yards of the shore. While whales don't always pass that close, the kelp beds that grow right off the rocks sometimes seem to draw them in. The first killer whale I saw swim through the kelp at Lime Kiln was J2 Granny. Video camera in hand, I wasn't looking through the viewfinder. Instead, my eyes were on the twenty-foot-long female killer whale rolling in the rope-like seaweed known as bull kelp about ten yards from me. Several stalks of kelp draped across her head and body. Her dorsal fin was almost completely obscured by the greenish-brown fronds, but the half-moon notch on the fin's trailing edge was still visible, which allowed me to identify her. Lifting her tail into the air with a piece of kelp held aloft in the notch between her tail flukes, Granny dove. A few seconds later, she surfaced, rolling sideways and lifting her pectoral fin into the air, waving more kelp like a flag in the summer breeze.

The whole episode probably lasted less than thirty seconds, but the moment is frozen in time for me. I recall it as clearly as if it happened yesterday. Any type of close encounter with a wild animal, let alone a top predator as large as an orca, is exhilarating and could be terrifying. Being next to an animal of Granny's size, though, I didn't feel afraid. Killer whales don't elicit the fearful response one might expect from other Pacific Northwest predators, such as a wolf or a bear. Part of that may have to do with being safely on land while the whales are in the water, but the lack of fear holds true for me even when observing orcas from a boat or kayak. What I experienced that day was wonder, drawn by the undeniable charisma of these sleek black and white animals with triangular dorsal fins.

The whale before me had been honed by fifty million years of evolution to be the master of her environment. The wolf-like ancestors of all cetaceans (the taxonomic name for the group made up of whales, dolphins, and porpoises) reentered the water, and over evolutionary time their bodies morphed to match the marine environment while still holding clues of their previous terrestrial life. The nostrils shifted to the top of the head, making it easier to replenish oxygen when breaking through the water's surface. The forelegs became large pectoral flippers to aid in directional swimming, the five sets of metacarpal bones (as in human fingers) still present. The hind legs all but disappeared, fused back into the body except for a single vestigial bone, resulting in a more torpedo-like body shape. Where the legs used to be, a powerful tail with two flukes evolved to provide propulsion, and the stocky musculature to support this form of movement. A thick layer of blubber insulates the vital organs against the frigid waters, and special physiological adaptations ensure oxygen flow is provided to all essential areas even during deep dives.

The black and white markings on Granny's skin gave her a useful countershading, to help her sneak up on prey. From above, her black back blended in with the ocean depths; from below, her white belly camouflaged her against the lighter sky. These markings were starkly broken up with her eye patch and a swirled white marking on her flank; these bold colorations likely helped her coordinate hunting activities with her family without using sound. Finally, the dorsal fin, like the keel of a sailboat, provided stability in any sea conditions. None of these specialized adaptations—for speed, for warmth, for agility in the water—took any thought or effort at all. But it wasn't just her ability to thrive in an alien underwater world that drew me in. Once I learned that this wasn't just an anonymous whale, but an individual—Granny—I was hooked.

Here was an animal that had been living in these same waters for nearly a hundred years, for that was her estimated age. She came not by herself but with her family and extended community, consisting of lifelong social bonds, unique traditions, and their own way of behaving and communicating unlike any other species on the planet. Here on these rocks, where an island meets the sea, her world touched mine: the human world and the orca world converge at Lime Kiln. This fact keeps drawing me back. With that single sighting of Granny, I glimpsed the lives of our intelligent, aquatic neighbors. After that close encounter, I wanted more. Not just to see the whales again, but to learn more about them. How I longed to ask Granny, *Who are you? What is your story?*

J2 Granny was one of the oldest known Southern Resident killer whales and could be easily identified by the half moon–shaped notch in her dorsal fin.

Whether or not she was really a hundred years old (an estimated age), Granny was a special whale who had undoubtedly lived through a lot in local waters. She was young back when fishermen and even the Navy would shoot at orcas or use them for target practice. She lived through the capture era, when she was netted up along with the rest of her pod while dozens of her family members were taken into captivity or killed during the process. She saw the local salmon runs decline over the decades; she watched the boom of the fishing fleet and its subsequent dwindling. She saw the human attitude shift from one of harmful misunderstanding to one of such fascination that she and her family experienced a legion of admirers who view them from whale-watching vessels or from shore. One has to wonder what she made of it all.

When I entered the world of the killer whale in the early 2000s, there was a lot to learn about them based on the pioneering work of those who came before me. It would have been easy to take for granted that Granny could be individually identified and her familial relationships known, but it was not always that way. Just a few decades earlier, all killer whales looked the same to our untrained eyes, but that changed thanks to the keen observations of regional scientists who helped discover more about not only the species in general but also the specific types of orcas found in the Pacific Northwest.

RESIDENTS, TRANSIENTS, AND OFFSHORES

The terms "killer whale" and "orca" refer to the same species, with the scientific designation *Orcinus orca*. The name "killer whale" is thought to have originated hundreds of years ago by sailors who saw them attacking and killing other whales and thus called them "whale killers," a phrase that eventually was reversed. Some people feel this misnomer gives the whales an unfairly ferocious name, especially because not all populations of killer whales actually kill other whales. There are advocates who prefer to call them "orcas," although the scientific name traces back to the Latin, meaning something like "whale from the underworld of the dead," which is not that much friendlier. Throughout the book I use the terms "orca" and "killer whale" interchangeably. Killer whales are one of the more than eighty species of cetaceans in the world. Cetaceans are divided into two groups: odontocetes (or toothed whales) and mysticetes (or baleen whales). Dolphins make up the largest group within the odontocetes, and killer whales, despite their name, are actually the largest member of the dolphin family.

Early killer whale observers noticed that not all the whales they watched behaved the same. By the late 1970s it was clear to researchers that there were two distinct groups of whales in the inland waters of Washington and British Columbia. Some whales traveled in larger groups of a dozen or several dozen animals and were mostly seen in the wider waterways. They were active at the surface, vocal underwater, and upon further observation, were eating fish. Other whales were in smaller groups of just a couple of animals, or they even traveled alone. Quieter, stealthier, and often observed in narrow channels and inlets, these whales followed closely along rocky shorelines. At times, they would erupt into action in hot pursuit of a seal or porpoise, which they would kill and eat. When they did vocalize, their calls were completely different from the fish-eating whales. The large groups of whales were the orcas encountered most often, during the summer months seen as frequently as on a daily basis. Because of this habit of seemingly staying in the same region, they were dubbed "residents." The smaller groups of whales were infrequently encountered—sometimes weeks, months, or even years would go by between sightings of a particular group. This type was called "transients," because they seemed to roam much further and more variably. "Fish-eating" and "mammal-eating" are really more accurate descriptors, as both populations roam over a wide range but also may stay in one area for a period of time. The terms "resident" and "transient" are still widely used, although there has been a movement to call transients Bigg's killer whales in honor of the late Michael Bigg (more about him below).

Early theories varied as to what the relationship was between residents and transients, with some researchers speculating that the transients must be outcasts or dispersed members of the more gregarious resident groups. Over time, observations of residents and transients began to reveal some trends. The two types of orcas never associated with one another, and no whale was seen to switch from one group to the other. Was it really possible for two different groups of the same species to have overlapping ranges but have nothing to do with one another? The difference in diet seemed to be the key. Having divided up their shared habitat based on food sources, with one group eating fish and the other eating mammals, they curbed territorial conflicts that may have otherwise occurred. These differences in diet influenced their socialization patterns. The fish-eating whales were able to thrive in large, noisy groups without any impact on their fish prey. The mammal-eating whales, however, had no such luxury. Their marine mammal prey was more tuned in to their acoustic environment than fish, thus making them far more aware of the presence of a potential predator and their echolocation clicks or tonal social calls. Large groups of transient whales were not nearly as successful in hunting, so when family groups got too large, they would break off into smaller groups in order to survive. Opportunistic feeders, their stealthy travel patterns aided these smaller groups in searching for and surprising their next mammalian meal.

In the late 1980s the differences between residents and transients began to be more clearly defined. Scientists noted physical differences between the two groups, including sub-tle variation in dorsal fin shape and overall size. Additional work confirmed what the researchers suspected: residents and transients were genetically distinct, meaning that not only were they behaving differently from one another but indeed the two populations were not interbreeding. By the late 1990s, prey-sampling confirmed the observed dietary differences between the two groups. Resident whales were found to eat more than twenty different fish species, and regional transients had a diet of six different marine mam-mal species. There was no overlap in prey. We now know that it's not just any type of fish resident whales are after: it's salmon. Although they will eat a wide variety of fish species in small quantities, more than 95 percent of fish taken are one of the salmon species. Not only that, resident whales exhibit a selective preference for Chinook salmon, which are the largest and fattiest of the local species, giving the

Transient killer whale T137A spyhops with a porpoise fetus in his mouth.

whales the most energetic bang for their buck. More than 70 percent of their diet is made up of Chinook, with chum salmon coming in second at about 20 percent of their diet.

It is unusual for two populations of the same species to inhabit the same area and neither compete nor interbreed. Most often, over evolutionary time, two animal species diverge due to being either fully or partially geographically isolated from one another. If you don't come into contact, you won't interbreed, and over time your genetic differences will increase because you're not sharing the same gene pool. With resident and transient orcas, however, there is potential speciation happening without any geographic isolation at all. When this so-called sympatric speciation is occurring, it means the gene flow between two populations is being interrupted by something other than geography. Often, this is due to a biological or ecological mechanism, but among resident and transient orcas, the reasons for not interbreeding seem to be cultural. They are choosing not to interact based on their different behavior.

There is enough variation between them that residents and transients are now considered distinct ecotypes: different forms of killer whale. Biologically, there's nothing to keep them from communicating, socializing, or breeding, yet they do none of the above. The cultural differences between the two groups seem to come with a rule that says you only associate with whales that are like you. The implications of this rule are something we are just becoming aware of, and the impact of this rule on the different populations of orcas could be profound. If given the choice between eating a marine mammal or going hungry, it seems a salmon-eating resident whale will starve. As prey studies have indicated, the rule seems to go even farther than that. The resident whales may settle for no less than Chinook salmon. When faced with abundant pink salmon during low Chinook salmonid years, the whales still seem, perhaps due to cultural rules, to refuse the pink and continue to search high and low for Chinook.

In the late 1980s killer whale researchers discovered that a third ecotype of killer whale inhabits the northeast region of the Pacific Ocean. Known as "offshores," these whales also travel in large groups, but as their name suggests, they tend to roam far offshore. In the decades since their discovery, comparatively little has been learned about offshore orcas, in large part because their pelagic habitat (the open ocean) leads to fewer observations than with more coastal whales. Very worn-down teeth were found on all stranded specimens, so it was theorized that the diet of offshore orcas was probably sharks, who have rough sandpaper-like skin. This was recently confirmed: prey samples have been primarily Pacific sleeper sharks but have also included other shark species. On a few occasions over the past forty years offshore orcas have roamed into inland waters, even being spotted near Victoria, British Columbia. But they never remain more than a day or two, and these encounters have been few and far between. Interestingly, genetic evidence indicates that offshore and resident killer whale ecotypes are much more closely related than either group is to transient orcas. The estimated divergence of transients from all other killer whale ecotypes worldwide (not just residents and offshores) is seven hundred thousand years ago, providing compelling evidence that transients should in fact be considered a separate species. Comparatively, residents and offshores diverged an estimated eighty thousand years ago, with data suggesting that resident orcas are probably offshoots from offshore populations.

The transient and resident orcas off the coast of Washington and British Columbia are among the most studied marine mammals in the world, but they are far from the only killer whale populations. Orcas are actually one of the most widespread species on the planet, inhabiting every ocean of the world. Everywhere orcas are found, they have specialized to feed on whatever is available. Off the coast of South America, for example, some killer whales eat sea lions, as filmed in dramatic footage in many nature documentaries. In the north Atlantic, one population of orcas specializes on herring. In Antarctica there are several different types of killer whales that feed variously on seals, minke whales, or other large whales. The "resident" and "transient" distinction is unique to the Northeast Pacific, but there are different types and subtypes of orcas all over the world. Some suggest that the differences between types of orcas are great enough that there should actually be multiple species of orca designated. For the time being, all orcas are considered members of the same global species, although scientists recognize it to be a "species complex."

RECOGNIZING WHALES

Imagine you walk into a busy grocery store in a small town you've never visited before, and your goal is to figure out the relationships between all the people in the town based solely on your observations at the store. You're free to return as many times as you like, but you can't talk to any of the people, only watch their interactions. On the first day, you might see a man and a woman, about the same age, with two small children in tow, and you might feel comfortable concluding that this is a family: a married couple and their two offspring. Another couple catches your eye, this time a teenage boy and girl. Are they brother and sister, boyfriend and girlfriend, or just friends? If you watch their interactions long enough, you might be able to deduce their relationship. If they hold hands, they must be a dating couple. If they show a resemblance to one another, perhaps they are siblings.

Other people might be more difficult to categorize. That man by himself, is he a bachelor? Or does he have a family waiting for him at home? Do the children and grandchildren of that elderly lady live here in this town, or have they moved away? Is that young woman the older sister of those children she is accompanying, or is she the babysitter? If you think in detail about how you would accomplish this task, you would have to spend a great deal of time in the grocery store. You would have to record a lot of observations to deduce the many and varied relationships between the people, and the longer you watched, the more some of your original assumptions might start to be questioned. In other words, the more you observe, the more nuanced your understanding of the varied relationships would become. This thought experiment gives you an idea of what the first whale researchers were up against when they set out to learn about the killer whales of Washington and British Columbia, except their task was even more difficult. By looking at people in the grocery store, you would have immediately been able to tell them apart by their appearance, particularly their faces. Without being able to use facial recognition to identify individual whales, whale researchers had to discover a different technique.

In 1971, Canada's Department of Fisheries and Oceans (DFO) hired marine mammal scientist Michael Bigg to conduct the first-ever orca population census. Born in England, Bigg moved at the age of eight with his family to Vancouver Island, a perfect environment to foster his lifelong love of nature and wildlife. After graduating from the University of British Columbia, he worked at the Pacific Biological Station in Nanaimo. His PhD work was on harbor seals, but he soon became enamored with killer whales. Bigg's task was a daunting one; whales spend so much of their time underwater, and a group seen one day might be observed many miles away the next. He ambitiously set out to do a one-day census throughout Washington and British Columbia, asking fishermen, ferry crews, pleasure boaters, lighthouse keepers, and anyone else with eyes on the water to report orca sightings for one day in July 1971. Of the more than fifteen thousand surveys he distributed, Bigg got about five hundred back, and the initial findings were a shock. He estimated there were between 200 and 350 killer whales in the region, a far cry from the many thousands that marine parks conducting live captures claimed were roaming Pacific waters.

Bigg repeated the census survey in 1972 with similar results. He also personally spent time in Johnstone Strait off northern Vancouver Island that summer, where many of his killer whale sightings came from. Photographing the whales on the water, Bigg and his colleagues noticed that certain individuals were distinct with scarred or notched dorsal fins. He could pick out these whales on consecutive days. Looking more closely at his photographs, Bigg realized he could see fainter scarring or more subtle unique markings on every whale. If every whale could be individually photoidentified by their natural markings, it would be possible to conduct much more in-depth field studies on killer whales than anyone had previously attempted. The potential to study life history of the killer whales was obvious to Bigg, but the idea of distinguishing individual animals in the wild was still a controversial one.

K1 Taku, who had two notches cut into the top of his dorsal fin, helped prove Michael Bigg's theory that killer whales could be individually identified. Photo by A. R. Hoelzel.

In 1973 the adult male orca K1 Taku was captured with his K-Pod family in Pedder by Sealand of the Pacific, an aquarium that housed killer whales near Victoria, BC While two of Taku's family members were removed for life in captivity, he was deemed too large to take and turned over to Bigg, who wanted to prove his new idea that killer whales could be individually identified. With the help of a veterinarian, Bigg cut two notches into the top of Taku's dorsal fin to test his theory that such marks were permanent and could be used to individually identify killer whales long term. In addition, Taku was fitted with a small radio transmitter, and Bigg hoped to be able to track Taku's movements for at least a month. Two and a half months after his initial capture, Taku was set free. Bigg followed Taku by boat, but unfortunately for the researchers, the transmission quality from the radio pack was poor, and they lost the whale after about eight hours. Taku was not seen again for nine months, but on Bigg's next sighting, Taku had reintegrated with the rest of K-Pod and no longer had the transmitter attached. The two distinct notches were unchanged, however, making Taku an easy whale to identify both then and for the rest of his life.

DFO wanted Bigg to refocus on his work with seals and sea lions, but killer whales were his true passion and he continued to study them in his spare time. Using his broad network of sightings contacts, Bigg regularly dropped everything over the next three years to go see killer whales, logging hundreds of whale encounters in British Columbia and Washington. By the end of 1975 he felt confident he had developed a photo-identification catalog of nearly all the orcas in the region, using the size and shape of their dorsal fins and the distinct gray marking that sits behind the fin called a saddle patch. Some saddle patches are solid gray (or "closed"), while others are checkmark-shaped or ("open"). Other saddle patches are wispy, have dips, or have protruding lines called "fingers," making each whale unique.

The "open" saddle patch of K20 Spock contrasts with the "solid" saddle patch of her son, K38 Comet.

With this discovery, researchers were confident that they were indeed recognizing the same whales again and again. But now that they knew they were seeing the same whales on repeated occasions, how could they determine the relationships between the whales? In the grocery store scenario, you would have the advantage of some preconceived notions of how human societies work. Men and women get married and have children in nuclear family units. People of a similar age often hang out together as friends. It's not unusual for someone other than a family member to watch the children.

A husband and wife may not always be seen together, but that doesn't mean they aren't still a monogamous couple. The killer whale researchers had no such basis to work from. Did orcas have nuclear family groups, like humans? Were groups made up of a dominant bull and a harem of females and their offspring, like with chimpanzees? Was there an alpha pair, like with wolves? They had to piece together a very complicated puzzle with little information.

In 1976 the National Oceanic and Atmospheric Association (NOAA) in the United States hired whale scientist Ken Balcomb to evaluate Michael Bigg's findings. Balcomb partnered with Bigg and dedicated the summer and fall to photo-identification efforts in Washington waters. He agreed that the resident whales he observed segregated into three distinct groups, fitting the pattern Bigg had noticed with another resident population of orcas off the north end of Vancouver Island, where they had sorted the orca society into pods—a term defined by them as a group of whales that spend more than 50 percent of their time together. Up north, Bigg had designated the first group A-Pod, then B-Pod, all the way to I-Pod. With this more southern population of whales, they decided to continue with that mnemonic, and the three groups in the Salish Sea region became known as J-Pod, K-Pod, and L-Pod. The pods from up north kept to a different range, with the dividing line in British Columbian inland waters occurring north of the Fraser River in the Strait of Georgia. These two separate orca communities were designated "Northern Residents" and "Southern Residents," respectively. Using Bigg's method of giving individual whales within each pod an alphanumeric designation, Ken

Balcomb and colleagues referred to the Southern Residents the same way: J1 was the first whale identified in J-Pod, J2 the second, and so on. The most iconic and recognizable whales were numbered first, and the fact that J1, K1, and L1 were all adult males perhaps reveals something about the preconceived notions of the researchers. By 1976 every resident whale was numbered and associated with a particular pod. At this time, there were more than two hundred whales in the Northern Resident community and seventy in the Southern Resident community.

Unlike any other species in the animal kingdom, male and female resident orcas stay with their mother for their entire lives. In all other animal species, one or both genders will disperse from their natal family group, but not so with resident orcas. When a group of killer whales was encountered by early researchers, the initial assumption had been that a large male in the group was probably the dominant bull. Closer observation revealed that this male was actually traveling right next to his mother, and he was surrounded by his brothers, sisters, nieces, and nephews. The pods identified by the researchers were extended family groups, made up of several related females and all of their offspring. A matriline was an immediate family consisting of a female, her offspring, her daughters' offspring, and so on—up to five living generations at a time. While pods spend most of their time together, they do occasionally split up into smaller groups. Matrilines proved to be even more stable, with immediate offspring always staying with their mother. Matrilines are designated by their oldest female, so for instance "the J16s" refers to the matriline made up of J16 Slick and all of her living descendants.

As Bigg and his colleagues learned about residents and transients and their various pods, they realized there are multiple levels of social hierarchy among killer whales. The species *Orcinus orca* is split into different ecotypes, and among ecotypes are different populations, also called communities. Within each community there can be from one to several acoustic clans, or groups that share vocalizations and call types. Among Southern Residents, J-, K-, and L-Pods are all part of J-Clan, while there are three acoustic clans among Northern Residents. Each acoustic clan is further made up of several pods, and each pod of several matrilines. Every individual whale thus belongs to a variety of social groups. For example, K20 Spock is of the resident ecotype, belongs to the Southern Resident community, is acoustically part of J-Clan, a member of K-Pod, and part of the K13 matriline. Harnessing this knowledge of resident orca life history, Bigg and Balcomb used the size of each whale to estimate age and the association patterns to establish probable familial relationships. They inferred the specifics of every pod and matriline, filling in new data of confirmed births and deaths as they observed them each year. They put this all together into the genealogy guides that are now so familiar to naturalists in the region, indicating the ages and relationships of every whale in the Southern Resident population. Every page of the identification guide reads like a family tree, showing the history of every matriline within the three pods.

A wrench was thrown into the Southern Resident killer whale identification system in 1977, however, just a year after the alphanumeric naming system had supposedly been finalized. A group of six whales that had been given

L-Pod numbers started spending more time with K-Pod. The familial relationship between these whales wasn't figured out with certainty, but they were always together, assumed to be a related matriline called the L18s. By 1981, though, the L18s had seemingly switched pods, spending all of their time with K-Pod. After learning their acoustic vocalizations were similar to K-Pod, Bigg thought it would make sense to "move" them. The researchers weren't quite sure how to handle this, so for the time being considered the whales part of L-Pod and waited to see what would happen. In the fall of 1986 a new calf was born to this L18 group. At the time, with the whales still traveling almost exclusively with K-Pod, the decision was made to switch the designation of this subgroup from L to K—the only time resident whales have had their alphanumeric designations changed. L18 became K18, L19 became K19, and so on. The timing was perfect, as switching the numbers from L to K wouldn't result in any duplicate numbered whales in K-Pod. To avoid confusion, they decided not to redesignate the L-Pod numbers they lost in the switch, so as a result there are six numbers that no longer occur in the L-Pod genealogy. The new calf, born to K18, was designated K21and named Cappuccino. The K18s provided the first clue that matrilines are more stable over time than pods. Other whales and groups of whales have seemingly shifted pod associations for both short and long lengths of time, but no other whales have had their alphanumeric designations reassigned.

The most iconic (and most photographed) whale in the Southern Resident community was undoubtedly J1. Nicknamed "Ruffles" for his wavy dorsal fin that stood five feet high, J1 was the most recognizable whale in the population. When I worked on a whale-watching boat, many return visitors remembered Ruffles and wanted to see him again. Others, first-time whale-watchers, had heard about him, and he was the one whale they wanted to see. Most regional whale-watchers have at least one good Ruffles story. My natural affinity was instead with J2 Granny, Ruffles's near-constant companion. He and Granny together made such an impression on me that I painted a mural of them in the houseboat where I first lived full-time on San Juan Island. Every time I wrote a blog post about the resident whales, I sat beneath the mural.

J1 Ruffles, named for his wavy dorsal fin, remains one of the most popular and iconic Southern Residents.

While I was lucky enough to have some close encounters with Ruffles over the years, my most lasting image of him is from far away. I picture him surfacing far offshore, his towering dorsal fin easily visible from the distance of a mile or more. He had a characteristic way of surfacing: slow, deliberate, with a strong thrust of his flukes as he dove that caused the last visible tip of his dorsal fin to lurch forward just before it disappeared. When the rest of J-Pod was tightly grouped or closer to shore, he might be off on his own, almost as if standing guard. Of course, Ruffles wasn't always by himself. He often hung out with the young males; I saw him in close association with every other J-Pod male of every age. The whale-watch community called Ruffles "the man," thinking of him as not only the most iconic but somehow also the most masculine, big, and tough of all the whales. This view was somewhat substantiated with the genetic paternity research done on the resident population. It turns out that Ruffles was the father of many whales, not only in K- and L-Pods but even in J-Pod.

At the one hundredth birthday party of Granny (held on July 2, 2011, at Lime Kiln Point State Park), Ken Balcomb explained how they arrived at 1911 as Granny's estimated birth year, a story I had never heard. Photos of both Ruffles and Granny existed in 1971, and both whales were already full-grown adults. Since orcas reach full size around age twenty, researchers made the estimated birth year for Ruffles as 1951 (1971 minus twenty years). Based on the way Granny and Ruffles associated with one another, they suspected that she might in fact be Ruffles's mother. Because Granny was never seen with a new calf since the study began, researchers assumed she was postreproductive and that perhaps Ruffles was her last calf. Females generally stop reproducing around age forty, so if Granny had

Ruffles when she was forty, her birth year would be about 1911 (1951, the birth year of Ruffles, minus forty years). This estimate was based on a lot of assumptions, some of which—thanks to additional studies and technology—we now know were inaccurate. For one, genetics indicated that Granny was not Ruffles's mother; in fact, genetic research by Mike Ford and colleagues would point to J1 likely being the offspring of L45, an older female from a completely different pod! In light of these revelations, Granny and Ruffles could have actually been much older (or in Granny's case, much younger) than estimated. The fact that we don't know only adds to the mystery of these iconic whales.

LEARN MORE

For further reading about different killer whale ecotypes, see *Transients* by John Ford and Graeme Ellis, *Killer Whales of the World* by Robin Baird, and the special issue of the *Whale Watcher*, the journal of the American Cetacean Society, titled "Killer Whale: The Top, Top Predator."

CHAPTER 2

Living in a Close-Knit Community

MUCH OF AN ORCA'S LIFE occurs underwater, where we can't easily see what's going on. Behavioral observations are therefore focused on what we can see when a whale breaks the water's surface. The most common behavior, if you can call it that, a whale-watcher will see is the whale coming to the surface for a breath. It's a simple behavior, but it is amazing how much misinformation some people have about this. For instance, on several occasions while working as a naturalist, I was asked why whales come to the surface. The first time, it took me a moment to realize that the answer the visitor was looking for was, "to breathe." It was remarkable to me, as a whale lover, that someone could be so unfamiliar as to not even know whales are air-breathing mammals like humans—a stark reminder of how disconnected we have become from our aquatic mammalian neighbors, and of how much education can happen during a whale-watching trip. Another inaccuracy I hear is when observers say, "Look at the blowholes!" The blowhole is literally the whale's nostril, which over evolutionary time has moved from the front of the face, as it is on land mammals, to the top of the head. Baleen whales have two blowholes, just like humans have two nostrils. In toothed whales, including orcas, the two nostrils have fused to form a single blowhole. What we are really seeing when a whale surfaces is not the blowhole but the blow (or spout), as the water on top of the blowhole gets blown into the air as a spray when the whale forcefully exhales.

Killer whales, when they surface, do so much more slowly and deliberately than most other dolphin species, who often seem to swim at hurried paces and burst through the water for just a split second. The surfacing of an orca is a dramatic event to witness in itself. First, the tip of the dorsal fin becomes visible, slowly rising to its full height above the water. Just before the fin has fully cleared the surface, the top of the head breaks the surface, and the whale subtly lifts its chin to make sure the blowhole on top of its head is completely out of the water. The blowhole opens, and the whale exhales the air that is left in its lungs with a powerful *kawoof!* This sound is as much a part of the true whale-watching experience as anything you will see. Immediately following the exhalation is the next inhalation—a quieter sound that is only audible in the most peaceful viewing conditions. Right after inhaling, the whale begins to sink back below the surface, often at the same measured pace. The head disappears, then the back, until only the dorsal fin is visible, slowly descending into the murky depths. Almost without a ripple, the tip of the dorsal fin vanishes, and the only sign of a whale having been there is a slick at the surface indicating where the water was disturbed.

Unlike humans, who are unconscious breathers, orcas have to consciously make each inhalation and exhalation. That means every breath is a deliberate act. John Lilly, the famous researcher who studied dolphin neurophysiology, discovered this when he sedated a dolphin for one of his studies. The dolphin, unable to consciously breathe, died. So how do orcas and other cetaceans sleep? Not at all like land mammals do! Because dolphins can never fully fall asleep, they have adapted a very different mechanism of resting wherein they only rest half their brain at a time. Even when "sleeping," dolphins continue swimming and surfacing, though at a much slower pace than when they are fully awake.

THE WAYS OF THE WHALES

The broad categories of behavior a group of whales engages in are traveling, foraging, resting, and socializing/playing. When Southern Residents rest, they usually get into a tight group and all dive and surface together, often staying below the surface for minutes longer than when engaged in any other type of behavior. Unlike Northern Residents, Southern Residents do not vocalize at all while resting. It's an amazing sight to see so many dorsal fins at the surface as a resting line of whales comes up for air: sometimes more than twenty fins will be visible at once. Then, they all dive together and the water is still for four to eight minutes until they all come up again. Southern Residents usually rest for a couple of hours at a time.

When traveling, the orcas head in a straight path from Point A to Point B, often moving along one of their typical traveling routes. All the whales swim in the same direction and travel at the same speed, usually a medium pace of

three to six miles an hour. The whales are in looser association with one another than when they are resting, although
often matrilineal family groups swim right together. This
means that every female is usually traveling with all of her
offspring flanking her, the youngest animals closest to her.
When watching J-Pod go by, for example, you might see the
J16s, followed by the J22s, then the J17s, and so on. Often
the different family groups are only separated by a matter
of a few hundred yards, but other times a whole pod traveling in the same direction might be spread out over several
miles. If orcas are traveling at high speeds (they can reach
top speeds of more than thirty miles an hour), they surface
in a behavior called "porpoising." They lunge much higher
out of the water than on a regular surfacing, creating a huge
white wash of spray on either side of them. Orcas can maintain porpoising speeds over distances of several miles.

Depending on what their main type of prey is, different
types of orcas undertake different behaviors when foraging. Transient orcas might quietly sneak up on their prey,
or they might engage in a dramatic group chase of potential prey with lots of leaping and lunging at the surface.
Other mammal-eating orcas off the coast of Argentina,
for instance, actually lunge up onto sandy beaches to grab
sea lions right from shore. In the North Atlantic, pods of
orcas cooperatively herd schools of herring. Resident killer
whales are fish eaters that for the most part pursue fish
individually. Foraging resident whales are very spread out,
often over an area of at least several miles. There are two
typical types of foraging: milling-foraging and travel-foraging. When milling, the whales are nondirectional, meaning
that an animal may face one direction on one surfacing and

in a different direction on the next. There is no coordinated direction of travel among the group when they are feeding in this manner. One place milling-foraging is often seen is off the south end of San Juan Island near Salmon and Hein Banks, where the whales may stay in this area for hours on end. When travel-foraging, the whales are spread out but all traveling steadily in the same direction. Individuals or small groups will opportunistically break off to pursue a fish while the rest of the group continues to move in the same coordinated direction. One location travel-foraging occurs is in Haro Strait. All the subgroups will be spread out as they travel north, stopping to feed as they find fish. Often, all the groups simultaneously turn around, and the whales continue to feed in the same fashion while everyone moves slowly south.

Unlike when resting or traveling, whales will often surface alone when foraging as they pursue a fish with no other animals in their immediate vicinity. After a fish is caught, however, prey-sharing will sometimes occur among family members. Whales will converge toward the same spot on the surface, circling around one another as they tear up and divide the prey one whale has caught. A study of Northern Residents determined that nearly three-quarters of confirmed predation events involved prey-sharing, and of these the vast majority of sharing was among whales of the same matriline. Adult females were the most likely age/sex class to share prey, generally with their offspring. The benefits of prey-sharing could help explain why sons and daughters both stay with their mothers for life. Interestingly, postreproductive females seem to continue to prey-share with adult sons, but less so with adult daughters. This may likely tie in with the survivorship of adult males, which so closely follows that of their mothers.

When the whales socialize, they are engaged in playful activity and seemingly just enjoying one another's company. There is a lot of intermingling among groups of whales, with matrilines and pods breaking up and intermixing with one another. There tends to be lots of surface activity when the whales are socializing, and this is also when sexual behavior and mating occurs. The most obvious sign to a human observer that something sexual is going on is when a male surfaces upside down, showing his long, pink penis fully extended from his genital slit. Nicknamed "sea snakes" or "pink floyds," orca penises can be up to six feet long. Like other dolphins, orcas have strong musculature around their genitals, giving them some maneuverability of their penis to help facilitate mating. (This also makes them capable of picking up objects with their penis, a behavior commonly observed among captive dolphins.)

Among orcas, sex is far from the private affair it usually is among humans. Dolphins are known for being highly sexual animals who routinely engage in sexual behavior even when it is not tied to reproduction. Because of this, it is claimed that only dolphins, humans, and a few other species of primates engage in sexual behavior for pleasure. The Southern Residents are no exception, where in true dolphin fashion it seems that anything goes. Age doesn't seem to play much of a role in whether an orca is sexual or not. Young whales are often in close proximity when adults are engaging in sexual behavior, and it is not unheard of to see sea snakes from subadult males as young as a couple of years old. Older females appear to remain sexually active even during their postreproductive years. They often engage in sexual behavior with younger adult males, perhaps providing some sort of training. The theory of female orcas providing sexual tutoring to young males is further strengthened by the fact that when juvenile males are seen with adult females other than their mothers, it is not uncommon to see them with erect penises. The gender of group members doesn't restrict sexual activities. At times, there are groups of adult males from different pods associating sexually with one another when there are no females in their immediate vicinity.

Little is known about how orcas choose mates. There are no obvious signs of competition among males for access to reproductive females. Whenever there is sexual dimorphism between males and females of a species (where the two genders exhibit different physical characteristics, as in the taller dorsal fin of male killer whales compared to females), it is usually something that has evolved for the purpose of sexual selection, suggesting that the larger tail flukes, pectoral fins, and dorsal fins of adult male orcas make them somehow more attractive as mates to selective females. A lot of mating behavior occurs when all three pods get together. The term "superpod" refers to a multipod aggregation of orcas; among the Southern Residents, the term is specifically applied when all members of all three pods are present and together. Superpod gatherings, especially when groups of whales initially come together, tend to be highly social. Sometimes, particularly in August and September, all three pods spend more extended periods together and may travel in one large group for days or even weeks at a time.

While resting, foraging, traveling, and socializing are the broad behavioral categories for whale groups, the behavior of the entire group or even an individual whale doesn't always neatly fall into one type of activity or another. A playful group of whales may be steadily traveling in one direction. While the pod is traveling, one animal might break off for a moment and mill around in an area in pursuit of a salmon. When most animals are foraging, another may hang at the surface for a minute or two in a behavior called logging, thought to be the whale equivalent of a nap. A peripheral group of whales that is engaged in play behavior while the rest of the group is foraging often contains one or more juvenile whales. It's not uncommon for one adult whale to babysit a group of rowdy juveniles while the rest of the pod rests or feeds.

Orcas also exhibit several common surface behaviors. One class of these behaviors is the fin slaps, where the whale will raise either its tail or pectoral fin above the water and bring it down to slap the surface, creating a spectacular crashing noise both above and below the water. Another less common behavior occurs when a surfacing whale rotates quickly, slapping its dorsal fin on the top of the water in a dorsal fin slap. Tail slaps are by far the most commonly observed surface behavior from Southern Residents. As with most orca behaviors, the meaning of actions like a tail slap aren't really known. Most surface behaviors seem to increase in frequency when the whales are socializing, suggesting that they might be play. At times, it really does look like goofing around, as a whale will be lazily swimming on its back with pec fins up in the air, slowly slapping its tail over and over. Other times, these behaviors are more aggressive.

On a few occasions I've seen whales tail slap so furiously that their tail is completely obscured in the constant splash they're creating. North Atlantic killer whales stun their herring prey via tail slaps, although it is unknown if Southern Residents ever do the same when hunting salmon. Since fin-slapping behaviors create a loud sound, they are also called percussive behaviors and may also play some communicative role. Another behavior that looks almost like an overenergetic tail slap is the cartwheel. When an orca cartwheels, the back half of its body lifts up out of the water and crashes down to the side.

The most impressive orca behavior also creates a tremendous splash: the breach. When breaching, a whale propels itself up through the surface until almost its entire body becomes briefly airborne. Most whale-watchers hope to see a whale breach, and the Southern Residents are a good group of whales to watch for that since they tend to breach more often than other orcas, such as the Northern Residents. For whatever reason, Southern Residents often get into a breach cycle, in which they breach multiple times in a row. This is a great thing for photographers to know! It's hard to be pointed in the right direction to perfectly capture a totally unexpected breach, but the pause after the resounding splash is a great time to get focused on the right spot in case the whale comes up and does it again. The meaning of breaching is purely speculative. Like fin slapping, the acoustic aspect of the behavior could be a form of communication, but it also looks like an expression of pure joy. Young whales, who tend to be a bit more rambunctious, are more likely to breach than adult whales. I once saw a young whale off Lime Kiln breach fifty-one times in a row as the rest of its pod traveled from north to south in comparatively stoic fashion. Adult whales do breach, however, including J2 Granny, even after her estimated one-hundredth birthday. Nothing is quite as impressive as seeing the entire bulk of a fully adult male launch into the air, although they seem the least likely to do so. During the ten years I watched J1 Ruffles, I saw him breach on just three different occasions.

J49 T'ilem I'nges (which means "singing grandchild" in the language of the Samish tribe) spyhops. Researchers think whales may do this to peer above the surface, but often their eyes are closed when they spyhop.

Imagining a breach from a whale's perspective is an interesting exercise. They live in a mostly dark, weightless environment, and during the brief action of breaching, they get to experience the sensation of gravity and, during the day at least, the full brightness of the sun. While fully adapted for their underwater life, orcas have maintained their ability to see both underwater and in the air.

There is one other behavior that gives them a glimpse of life above the surface: when a whale spyhops, it lifts its head and forward part of its body straight up out of the water and sinks straight back down. Again, we can only guess why the whales do this; a likely theory is to see above the surface, but often their eyes will be closed when they spy-hop, further adding to the mystery. Antarctic killer whales, for example, will spyhop by icebergs to get a look at any seal prey that may be taking refuge there. We can only imagine what Southern Residents are doing when they spyhop.

GROWING UP IN KILLER WHALE SOCIETY

The development and life span of wild orcas is remarkably similar to that of humans, although their size and external appearance couldn't be more different. Calves are born at six to eight feet long and weigh about four hundred pounds. They look like miniature versions of adult whales, except their saddle patch markings are usually very faint at first, and their white markings have a pink or orange hue. People offer varying reasons for this different coloration at birth. The most likely cause is that there's a buildup of red blood cells in the calves that gradually balances out as the liver begins to fully function; it's the same reason that many human infants appear jaundiced. In any case, this color-ation fades to white over the first six months or so. Calves nurse for the first couple years of their lives. At this age,

males and females can only be distinguished visibly by the
black-and-white pattern on their undersides, making it difficult for researchers to determine the gender of offspring
until they get older or until someone gets a photograph of
the whale's belly. Male orcas have a more elongated white
patch below their navel with a single black mark at the genital slit. Females have a shorter, rounder white patch, but
three visible black markings indicating the genital slit as
well as two mammary glands.

Orcas are considered juveniles until about six years of
age, and they are subadults until their teenage years, when
they begin to reach sexual maturity. During this adolescent phase, males and females experience different growth
patterns, resulting in the sexual dimorphism apparent in
adult killer whales. Females have curved dorsal fins that
grow up to three feet in height, and they reach lengths
of about eighteen or twenty feet. Males grow several feet
longer, have larger pectoral fins and tail flukes, and (most
obviously to humans) have a dorsal fin "sprout" where
over several years their fin grows from the smaller fin of
the juvenile and female to the straight, more triangular fin
distinct of adult male orcas that can be up to six feet tall.
Around age fifteen, females usually give birth to their first
calf. Calves have been born to mothers as young as nine,
and other whales have reached their early twenties before
being seen with their first calf (although it is possible some
presumed first-time mothers have lost a previous calf
that went undocumented, for example through a miscarriage). For Southern Residents, birth can occur at any time
of year, although more young tend to be born in the fall
and winter. The gestation period is a whopping seventeen

months. It is believed that cetaceans can only successfully give birth to a single calf at a time. Live twins have never been confirmed in orcas.

Not much is known about the birthing process in wild orcas. Usually there is simply no calf, then in the next observation of the mother, a new baby. There was one instance in 1990, however, where a Southern Resident was actually observed giving birth, documented in the journal *Marine Mammal Science*. When the researchers arrived on scene with the L4 matriline, the whales were spending a lot of time at the surface milling around in a tight area. One whale spun repeatedly at the surface in a behavior unlike any other they had witnessed before. After this thirty-second bout of spinning, a newborn whale appeared, pushed above the water on the rostrums of three other whales. These were family members of the mother, presumably pushing the calf to the surface for its first breath, a behavior seen in other dolphin species. Following this, all the whales exhibited lots of percussive behavior at the surface, repeatedly coming into contact with the calf and at times even tail-slapping on or near the newborn and pushing it out of the water with their bodies. The role of this somewhat aggressive behavior is unknown, although similar observations have been made during the births of other cetacean species. After twenty minutes of this, the whales began a long period of directional high-speed swimming with the newborn between its mother, L55 Nugget, and grandmother, L27 Ophelia. All of these behaviors—from the rotating at the surface to the calf being pushed by other whales to the percussive behaviors—were also observed in the only other detailed documented birth of a resident whale to a Northern Resident pod in Johnstone Strait. The newborn Southern Resident in 1990 was designated L82 and later named Kasatka. She was the firstborn offspring of Nugget.

Similarly, not much is known about nursing, since it occurs underwater and is rarely observed in the wild. How does a calf suckle a liquid while swimming in a liquid? There are several challenges to overcome. In cetaceans the nipples are concealed within the mother's mammary slits to help her keep her hydrodynamic shape when swimming. They can push the nipple out so it extrudes from the mammary slit, perhaps stimulated by nudging of the baby. But this is only part of the problem. Without lips or cheeks the calf can't create a suction seal around the nipple the way other mammalian babies do. Instead, they use their tongue as a straw. Baby dolphins have fringes around the edge of their tongue that interlock when they curl their tongue into a tube around the nipple. Instead of sucking the milk out of the nipple, mothers may actually shoot it out into the mouth of their young. It also helps that the fat-rich milk is much thicker than the surrounding seawater.

During the first year of their lives, newborn orcas are seen almost exclusively with their mothers. As they get older, they begin to associate more with other members of their pod, although the bond with mom always stays strong. Juveniles especially spend a lot of time with their older siblings and grandmothers; babysitting is common among resident killer whales. The amount of time a young orca spends with its mother is proportionate both to its age and the number of siblings it has. Juveniles with younger siblings spend more time with other whales, while those with no younger siblings spend more time with their mothers.

A hallmark of resident killer whale society is a lifelong connection with the maternal family group. The calf, K42 Kelp, is surrounded by his mother and brother as well as by J8 Spieden, an elder J-Pod female known for babysitting youngsters.

While any whale may take a shift watching over a boisterous young calf, some whales are especially good babysitters, such as J8 Spieden. Her estimated birth year was 1933, so she was assumed to be a postreproductive female when studies began in the early 1970s. While Spieden's name came to denote a local island, she was originally known as "Auntie Spidey" for a spider-shaped scratch on her saddle patch. She was easy to identify due to her relatively short and stubby dorsal fin. She was also the only Southern Resident you could readily identify without seeing: in calm conditions you can easily hear whales exhale as they come to the surface to breathe, but Spieden had a distinct, wheezy inhalation that sounded like air blowing over the top of an empty bottle. Spieden was the only whale I could identify in the fog! In early years, researchers observed her spending a lot of time with J4 Mama's offspring, particularly the older juveniles when Mama had a new calf in tow. Despite no longer having calves of her own, Spieden was helping to care for what may have been her grandchildren. In later years, her caring behavior extended beyond her immediate family members. She spent a lot of time with orphaned juvenile J32 Rhapsody in the late 1990s after her mother died. Also during this time, Spieden was often seen with J33 Keet, who was getting less attention from his mother after the arrival of younger sister J36 Alki.

If their young survive, killer whale mothers have a calf an average of once every five years, although the calving interval ranges from two years (if the calf doesn't make it) to more than ten years. The long period between birthing young is a result of the huge biological investment in every offspring: after the eighteen-month gestation period there are several years of nursing. A reproductively successful female may have just five or six offspring over the course of her breeding years. Males reach sexual maturity around eighteen or twenty, and will breed for the rest of their adult lives. The average life span of male orcas is about thirty years old, although some live much longer. Studies have shown older males sire more offspring than younger males. At the end of her breeding years, in her early forties, an adult female orca will experience menopause. In the wild, animals usually die shortly after their breeding years end; genetically speaking, they are no longer contributing to future generations. This is not the case with killer whales, however. After ending their reproductive years, older female orcas may live another decade or more, up to and even surpassing one hundred years of age. Clearly, they are still playing an important role to their pods, and as we've learned more about killer whale society, we've been given a glimpse into what this role might be.

The fact that resident whales live in a matrilineal society—with social groups based on each whales' relationship with its mother—provides a clue to the role that older, postreproductive females may be playing. One theory is known as the "grandmother hypothesis," where older, nonreproducing females live to help extend the survival of their offspring and grandoffspring. Grandoffspring survival doesn't seem to be affected by having a living grandmother, but having a living mother *does* affect the survival of her children, particularly adult males. A recent study showed that adult males over age thirty are fourteen times more likely to die within a year after their mother's death. Observations have shown that it seems necessary for a male, regardless of age, to find a replacement mother figure if his own mother dies. Often an orphaned male will latch onto a related female as a surrogate mother.

The most extreme example is the case of L87 Onyx. As a member of the L12 subpod, Onyx had several options for a new mother figure within his group after the 2005 death of his own mother, L32 Olympia. Instead, he surprised everyone by starting to permanently travel with K-Pod and their matriarchs, K7 Lummi and K11 Georgia. By 2010, these elder females had passed away, and in the middle of the year Onyx switched pods again, moving to J-Pod and traveling with that pod's elder female members, J8 Spieden and J2 Granny. (With J1 Ruffles passing away that same year, and it being revealed later via genetic analysis that apparently he was also an L-Pod whale who permanently traveled with

J-Pod, the mystery behind Onyx's movements has since deepened even further.) We used to think a killer whale would never switch pods. In the process of doing whatever is needed to survive, Onyx led observers to question everything we thought we knew about killer whale social structure. As is often the case, just when we think we've figured something out, the whales do something different.

In addition to the grandmother hypothesis, the oldest females may be the matriarchs, the leaders of the matriline or even the entire pod. It's often said that resident killer whales are matriarchal (meaning that elder females have power and influence); while this is likely true, we don't really know for sure how a pod makes decisions. The elder females undoubtedly carry a lot of cultural knowledge for the group about traveling routes, food sources, and the places to be in different seasons under different conditions. It isn't unusual to see an older female out in front leading a traveling pod, for example, and studies have shown they lead even more often in years of low food abundance, supporting the theory that elder females know where to go. The same older females Onyx latched onto were known for leading their pods: Granny and Spieden were often a mile or more in the lead for J-Pod, swimming steadily onward. In K-Pod, Lummi with her twin notches and Georgia with her checkmark-shaped saddle patches could be seen playing a similar role, although Lummi in particular seemed to have more of a relaxed attitude about her. She was known for taking a lot of catnaps, where she would spend minutes logging at the surface. One researcher recalled heading in the direction of a whale sighting when they came across a single orca hanging at the surface. It was Lummi, and as the research boat approached, it seemed to wake her up. She twitched back to life and started swimming toward where all the other whales were, by this point a couple of miles away. Perhaps this was just a personality quirk of Lummi's, as in general older females show no signs of their age. They are as active and swim as fast and as far as any other member of their pod.

These older females are the glue that holds a family group together. While no official studies have been done, pods sometimes alter travel and social patterns after the death of an older female. For instance, researchers anecdotally noted that J-Pod switched their primary travel route to the Fraser River from Boundary Pass to Swanson Channel after the death of one matriarch. Another example is that K-Pod started fractioning into smaller groups more often after the deaths of K7 Lummi and K11 Georgia. Elder females are central to pod associations, continuing to spend a lot of time with their siblings and adult offspring as well as the descendants of their adult daughters. Currently, many of the elder females identified when studies began in the mid-1970s are dying. A large part of the next generation of whales was removed from the population during the live capture era for display in marine aquaria, so it remains to be seen if or how comparatively younger adult females will begin to fill in these matriarchal roles. With the death of J2 Granny in 2016, J19 Shachi was the next to play the role of leader of J-Pod at less than forty years old.

As for adult male orcas, they tend to have fewer associations with other pod members than females and juveniles, becoming somewhat more peripheral group members; they are seen most often with their mothers or other males. While they undoubtedly are almost always within acoustic contact with their matrilineal group, they regularly forage alone at the edge of the group and often travel up to a quarter-mile or more away from other whales. As indicated by the decline of survivorship with the death of their mothers, it seems that adult males continue to rely on their mothers for inclusion in the pod. Instead of being dominant males with a harem of females, as early researchers originally suspected, the role of adult males seems to be the opposite: they might be subordinate in resident killer whale society.

Interestingly, outside of their family groups, adolescent and adult males' strongest associations are with other adolescent and adult males. It is not uncommon for groups of two to six males from different family groups and even different pods to travel together. Among mammals, male-male interactions are often related to maintaining dominance hierarchies; in other cases, all-male groups form to provide protection from predators (like bachelor groups of some ungulates, or grazing animals) or to increase access to mates (as in male bottlenose dolphin coalitions). Any nonaggressive social groups between males of the same species are usually made up of related animals. A thesis study done by Naomi Rose on these male-only social associations among Northern Residents provided some interesting insights to the role they might play in resident killer whale society. Male-only interactions almost always involve whales of mixed ages, and hence mixed sizes. This fact, along with the observation that male-male social groups are voluntary (no member is anxious to get away) and last for up to several hours, indicates that they are not playing a role in dominance determination among males. Indeed, no overtly aggressive behaviors are observed between the males. On the contrary, there is often sexual behavior occurring, with penile erections and lots of tactile contact between animals. This suggests that the males are coming together for social or play reasons, or perhaps to train younger males in rituals related to courtship behavior. The reasons for joining into these male groups may vary with age. Adolescent animals might be learning about appropriate sexual behavior from adult males, young adults may be burning off libido among peers since they have limited access to reproductive females, and adult males may simply be enjoying a close social interaction they don't often get in their now more independent life styles.

When socializing, males from different groups often seek each other out. J27 Blackberry swims with L95 Nigel during a multipod aggregation.

For the first forty years of study, no violently aggressive behavior was documented among killer whales at all, including among males trying to access mates. But just like

Onyx proving wrong the fact that killer whales won't ever switch pods, researchers in 2018 published a rare encounter proving wrong our assumptions about aggression in killer whales. An adult male transient killer whale and his mother were seen killing the newborn calf from an unrelated family group, the first documented case of infanticide in killer whales.

When a group of males does get together to socialize, it's nearly always memorable. One morning while working as a naturalist, I encountered such a group. All of the adult whales that were present were together: L73 Flash, L74 Saanich, L84 Nyssa, L41 Mega, and K21 Cappuccino. They were hanging out at the surface a lot, rolling around each other. Cappuccino pushed his head into the side of another male who was upside down with his pec fins in the air. At one point, Mega was lifted up out of the water by another whale underneath him. Another male, unidentifiable because he was upside down, surfaced with a fully erect pink penis. As we followed the whales up the strait, a freighter came by. I thought it might break up the whales' party, and I was right. As it chugged by, the whales spread out and dove. We saw the swell of the freighter wake head right toward where the whales had been. Suddenly I felt a surge of excitement. Were they going to surf? It was a behavior I had heard about but hadn't seen before: orcas surfing in a freighter wake, playing in a way similar to the Dall's porpoise that would sometimes ride our bow.

We got our answer a moment later. A tall dorsal fin broke halfway through the surface. It was Mega. He rocketed through the water, pushed both by his tail and by the wave, his fin creating a massive rooster tail splash that shot ten feet up into the air. In the next swell we saw the blurry black-and-white outline of another male zipping forward. All of a sudden, Flash lunged through the surface, almost completely clearing the water in a dolphin leap before disappearing behind the huge white wall of water of his own making. I'll never forget the sight of so many massive bulls lunging out of the water side by side.

Male resident orcas clearly enjoy each other's company, but how do they find mates? The offspring of most species disperse from their family group when they reach adulthood for one simple reason: mating opportunities. For the sake of genetic diversity, it makes sense to leave the family you were raised in to breed with unrelated individuals. Since male and female offspring of resident orcas stay within their maternal group for their entire lives, how do they encounter potential mates? Resident orca families regularly associate with other family groups within their community of whales. While Southern Resident orcas don't socialize with Northern Resident or transient orcas, J-Pod whales will often meet up and travel with K-Pod and L-Pod whales. During these multipod gatherings is when most sexual behavior is observed. A J-Pod male may go off and mate with a K-Pod female when their family groups are traveling together. When they split off again, he will go with his mother's group in J-Pod, and she will go with her mother's group in K-Pod. If a calf is born, it will stay with its mother and grandmother in K-Pod and have little obvious association with its father. It's possible that adult males meet potential mates through their mothers, gaining access to reproductive females that are the daughters of their mother's strongest associates. Researchers theorized that to maximize genetic diversity, females would only mate with males from other pods, and this was confirmed genetically in the Northern Resident killer whale population. However, the Southern Residents, perhaps because of their small population, do mate both within and between pods.

One result of the matrilineal social structure of resident killer whales is that mothers and daughters may breed at the same time, leading to potential reproductive conflict within a family group. A new study has shown that older females bear more of this burden, leading to a higher mortality rate of calves born to older females compared with those born to younger females. Thus, the unique social structure of resident killer whales may in fact have led to the evolution of menopause, which is so rare in the animal kingdom.

While the short-term associates of a particular whale can change throughout the day, it used to be thought that the long-term association of the pod was a stable one. L-Pod, being the largest of the three pods, has from the beginning of research in the 1970s been the least cohesive, regularly splitting into varying subgroups with anywhere from two to more than twenty animals. For the first thirty to forty years of observation, J-Pod and K-Pod were usually seen in their entirety, with all pod members traveling within a few miles of one another. Throughout the 2000s, however, it has become more common for Js and Ks to split into smaller groups. For instance, the J14s, the J19s, and often the J16s became known as J-Pod "Group A" while the J11s, the J17s, and the J22s became known as "Group B." This may be due to a decrease in food supply, requiring smaller foraging groups. In any case, this phenomenon has very much changed who whale-watchers and whale researchers regularly encounter. In 2000, for example, if someone reported J- and K-Pods traveling together, it meant all whales from both pods were present. Ten years later, if someone saw Js and Ks, it could easily mean that only half of J-Pod was there, perhaps with a single K-Pod matriline. It now seems that the static social unit among resident orcas is the matriline rather than the pod; even when pods split into smaller groups and travel alone or with other subgroups, matrilines of immediate family members virtually always stay together.

WHEN A WHALE ENDS UP ALONE

Killer whales, particularly residents, have lifelong social bonds that are central to their survival and well-being. But what happens when one of these social creatures ends up alone? Three stories with three very different outcomes provide a glimpse into what can happen when social bonds break down. In the fall of 1999, L67 Splash was seen with a newborn calf. Oddly enough, over the next few days that same calf was traveling with K18 Kiska. There was confusion as to who the calf's actual mother was, and researchers speculated they were perhaps witnessing an attempted "calf-napping." Kiska had recently been seen with a stillborn calf, and perhaps her mothering instinct was leading her to try to steal Splash's calf as a replacement. Over the next few weeks, both Splash and Kiska and their respective families were rarely far from the new calf, as he regularly went back and forth between them. Eventually the calf stayed with Splash and was designated L98. The following year, the calf was named Luna.

In the summer of 2001, when L-Pod returned to inland waters, Luna was not with them and was presumed dead. Also missing was Luna's uncle L39 Orcan. Nearly every time a whale is missing from their family group, it is because they have died. Luna proved to be the one exception to this assumption. A month later, a small killer whale was observed on his own in Nootka Sound, off the west side of Vancouver Island, BC. His unique markings helped researchers identify him as Luna and, strangely for a Southern Resident, especially a young one, he was alone. Luna appeared to be foraging successfully for himself, and because he looked to be in good health, the Department of Fisheries and Oceans in Canada decided to leave him be to see if he would reunite with his family over the winter. The Southern Residents can range off the west side of Vancouver Island, so maybe Luna would find his family on his own.

How did Luna become separated from his family, especially since his mother, Splash, was still alive? We don't know, but one potential clue may be the fact that Luna's uncle Orcan went missing over the same winter. If Orcan was babysitting Luna when he died, it's possible that Luna was left on his own, unable to relocate his family group. Others have theorized that there was something wrong with Luna or in the bond with his mother, as he regularly swam farther away from her than most other newborn calves do from their moms. Maybe Splash had been trying to get rid of Luna when he was born, rather than having him stolen by Kiska. Maybe Luna's family, for an inexplicable reason, was kicking him out or left him behind on purpose. We will never know for sure.

Meanwhile, word got out about Luna. In the absence of his whale family, this highly social youngster sought out companionship wherever he could get it. His orca vocalizations repeatedly went unanswered, and in a remarkable case of vocal learning, Luna began to mimic the barks of the sea lions who resided near him in Nootka Sound. Luna also started interacting with humans. He approached boats of all sorts to seek out interaction, and he loved to play and be touched. People feared the unregulated interactions were dangerous for Luna, and as attempts were made to keep people away, Luna became more aggressive and started damaging boats and other property in his quest for attention. Both scientists and some members of the public wanted to make an attempt to reunite Luna with his family. The local First Nations tribe believed he should be left alone, however, and other members of the public were also opposed. They believed an unsuccessful reunion would land Luna in captivity. Over the next four years the governments of the United States and Canada, the Mowachaht/Muchalaht tribe, and killer whale scientists failed to agree on what would be best for Luna. One serious attempt was made to capture Luna for a relocation effort, but after several days of on-the-water drama with different groups trying to lure Luna one way or the other, he was never captured.

In March 2006, an ocean-going tugboat took refuge in Nootka Sound during a storm. Luna approached it, as he did so many other boats. He got too close to the powerful engines, unfortunately, and was pulled into the propellers and killed. We will never know if Luna would have successfully reintegrated into his pod, but his legacy is nonetheless a lasting one that raised more questions than it answered. Why did he become separated from his pod? Would they have taken him back? Was it right to try to keep people away from Luna, when he seemed desperate for social interaction? Is it possible for humans to befriend a whale, and keep both parties safe?

Luna's behavior raised a lot of questions about the safest way for human-whale interactions to occur, if they should at all. Photo from the Luna Stewardship Project, provided courtesy of Kari Koski.

A similar situation played out in Puget Sound in 2002 when A73 Springer, a female member of the Northern Resident community, was found alone. There were several key differences that led to a different outcome for Springer. For one, she was actually an orphan, as her mother had died the previous year, making it perhaps more understandable how Springer might have become separated from her pod. Second, and more important, Springer was not successfully feeding herself and was in poor health. This led to quicker action among scientists to intervene, and while there was still a heated public debate, it was only a matter of months before a plan was put into action. Springer was captured in a net pen in Puget Sound and received a month of rehabilitation (with minimal human contact) to bring her back to health. She was then transported by boat to another net pen in Johnstone Strait in British Columbia, the core region of her family's home range.

While the relocation team initially intended to hold her for a while, the very next day members of Springer's extended family swam by and she could be heard vocally interacting with them. The scientists released her, and while she immediately swam toward the other orcas, she did not instantly reintegrate with her pod. Over the coming weeks, however, they seemed to accept her back, and Springer found a surrogate mother figure in A51 Nodales. This distant relative was seen directing Springer away from boats, a habit she picked up while in Puget Sound. The operation was declared a complete success when Springer returned with her family the following summer. Since then, she has become a mother, giving birth to a calf in 2013 and again in 2017. The success of Springer's rehabilitation and release was a stark contrast to the political tug-of-war that led to Luna's eventual death.

The final story of a wayward young whale begins in July 2005, when a report came in of a lone orca seen outside of Friday Harbor. Right away it seemed like an unusual sighting; this was a female or a juvenile, and the only time orcas are regularly seen on their own is when it is a lone bull transient. Regardless, early speculations were that this was a female transient until later in the day, when this whale made its way down San Juan Channel and out into the Straits, where the L12 subgroup was foraging. As the orca began trailing behind these whales, researchers took a closer look and discovered it was actually K31 Tatoosh, a six-year-old male from K-Pod, offspring of K12 Sequim. Tatoosh hung out near the L12s for the next three days, during which he was recorded making vocalizations shared by both K- and L-Pods. He looked to be in decent health, but where was the rest of his family? Still a juvenile, it was especially odd for him to be away from his mother.

When I heard a few days later that K-Pod was heading back to inland waters, my hopes rose that a happy reunion was imminent. The next morning, I was scheduled to volunteer with the Soundwatch Boater Education Program. When I arrived at the dock, the report was optimistic: not only had K-Pod arrived, we had a full superpod in Haro Strait heading north. It was a beautiful summer morning, so we were busy intercepting recreational boaters and handing out guidelines on how to boat safely around the whales. We didn't see K-Pod, but based on what we heard over the marine radio, Tatoosh was around but not traveling right with his family group. By midafternoon the superpod turned around, heading back south and offshore. As we moved further away from the island into the middle of the strait, the boat traffic thinned out. We found ourselves with no boats and seemingly no whales in the immediate vicinity. Slowly motoring south to catch up with the whales, we heard the distinct sound of a whale breathing and spotted a dorsal fin a few hundred yards in front of us. As we got closer and could see the saddle patch, our suspicions were confirmed: it was Tatoosh, trailing a half-mile or more behind the other whales.

What struck me as we neared Tatoosh was how small he still was at six years old. He probably wasn't undersized for his age, but something was wrong—this little whale shouldn't be on his own. He surfaced in a regular pattern, breathing three or four times then going on a slightly longer dive, and his breaths seemed full and strong. He seemed to have plenty of weight on him, with no sign of "peanut-head"—a visible depression behind the blowhole on malnourished whales. There were no obvious wounds or injuries on his body. After ten or fifteen minutes of traveling alongside him, we slowed down, preparing to head back in for the day. We bobbed in the gentle waves as Tatoosh continued heading southwest, following in the distant wake of the rest of his family. A few days later, he was gone, never to be seen again.

Often when a resident whale dies, we have little or no indication ahead of time that the whale is ill or something is wrong. Tatoosh was an interesting exception—he was clearly behaving differently for a period of days or weeks before his disappearance, but researchers had no other clues about what was wrong with him. What seemed especially odd to me is how he appeared to be shunned by his family. There are lots of stories of dolphins (and orcas in particular) aiding sick or injured family members, helping to keep them at the surface so they can breathe or carrying around their bodies after they have died. For Tatoosh, in his last days, although his family was in the vicinity, he was very much alone. This strange behavior by both Tatoosh and his family raised more questions for me, deepening the mystery of Southern Resident killer whale society.

LEARN MORE

For further reading about killer whale social behavior, see *Behavioral Biology of Killer Whales*, edited by Barbara Kirkevold and Joan Lockard, as well as scientific publications by Michael Bigg, Darren Croft, Emma Foster, and Brianna Wright. For more on Luna's amazing story, see the documentary *The Whale*, directed by Michael Parfit and Suzanne Chisholm. Other sources are *The Lost Whale: The True Story of the Orca Named Luna* by Michael Parfit and Suzanne Chisholm, and *Operation Orca: Springer, Luna, and the Struggle to Save West Coast Killer Whales* by Daniel Francis and Gil Hewlett.

CHAPTER 3

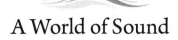

A World of Sound

MY INTRODUCTION INTO THE WORLD of killer whale acoustics took place in the summer of 2002. As a sixteen-year-old high school student, I had the chance to spend a couple of weeks with Bob Otis and his research team at the Lime Kiln Lighthouse, overlooking Haro Strait on the west side of San Juan Island. Bob and his team take behavioral data every time Southern Residents pass by the lighthouse between late May and early August. Taking full advantage of the opportunity to immerse myself in the whale world, I also spent time with Lisa, an intern for The Whale Museum, who was making acoustic recordings of the Southern Residents at Lime Kiln over hydrophones (underwater microphones). I sat alongside her as she identified different call types, and she showed me an acoustic catalog developed by Canadian biologist John Ford—the standard reference for call type classification in the region. I had spent a couple of evenings with Lisa at the lighthouse helping her put together an acoustics CD. She invited me to join her the next evening at the museum, where a group of interested individuals was getting together to talk about their current research and Southern Resident acoustics. Intrigued by what I had learned thus far, I accepted.

Assembled in a circle of chairs in the museum's upstairs gallery, the gathering commenced, with more than one person curiously eying the high school student in their midst (undoubtedly in part because I had bright blue hair at the time). Someone set up their computer and played some sample calls that the group identified to hone their skills at differentiating similar call types. For the Southern Residents each call is designated with the letter S, followed by a number. The call types aren't ranked in any way; they are simply numbered in the order identified: S1, S2, S3, and so on. Some call types have two or three different subtypes, indicated with i, ii, or iii after their number. After a particular call, the group had come to a consensus as to its ID. This was a call that Lisa and I had listened to repeatedly the day before, and I disagreed. Hesitating briefly, I spoke up. "I think it's an S2iii."

Everyone turned to look at me. The person at the computer queued up a sample S2iii (said "S2 type three") and played it for the group. It was undoubtedly the same call, and my ID had been correct. One man, bearded and several decades my senior, reached over the computer and played another call clip. "How about that one?" he asked.

"S6," I replied, relieved that it was one of the few calls I knew. The man said nothing but his eyes asked, *Who is this girl?*

After the meeting the bearded man introduced himself, offering his hand. "My name is Rich Osborne," he said. "I'm the research director here. Your identification skills are impressive. I think you should apply for an internship next year." I did just that, and spent the next five summers as a research intern at Lime Kiln Lighthouse for Osborne and The Whale Museum.

WHEN HEARING IS MORE IMPORTANT THAN SEEING

Whales aren't the only creatures to communicate acoustically; humans and most other animals do it to varying extent. People have alphabets, words, and rules of grammar that combine into advanced languages. But humans are also intensely visual creatures. We use our sense of sight to navigate through the world, to categorize and interpret things around us, and to read body language off of one another. Our acoustic sense is secondary to our visual one. For killer whales, it's the opposite. While they have fairly good eyesight, in their world visibility is poor. They may be able to see in detail a rock, a fish, or another whale few yards away, but how do they locate and chase down a Chinook salmon? How do they find their way through the complex waterways of the inland waters of Washington and British Columbia? How do they keep in touch with one another across not only yards but miles? The answer is sound.

All cetaceans evolved to use sound in a manner that humans will never be able to fully comprehend. Mysticetes (large baleen whales) reverberate with low-frequency sounds that can span entire ocean basins. Odontocetes (toothed whales and dolphins) use echolocation to determine their surroundings, bouncing sound off of objects and listening to echoes to create sophisticated acoustic maps of the surrounding world. Through echolocation, dolphins detect the size, shape, and distance of objects. Bottlenose dolphins, in an experimental setting, have shown the ability to detect a one-inch metal target at a distance of more than two hundred feet using echolocation. Echolocation can also provide odontocetes with internal information about an object. For example, they can differentiate species of fish based on the echolocation signal they get off the fish's internal air bladder, and they can use echolocation like a human ultrasound to determine things like pregnancy in other animals.

Sound in odontocetes functions not only as a navigational tool but also as a method of social communication. Sperm whales have codas, or patterns of clicks that form diverse call types. Bottlenose dolphins communicate primarily through various whistles, and each individual has a signature whistle that may function like a name. Orcas make three types of vocalizations. Echolocation, or clicks, are used for navigation and foraging. Whistles are simple pure tone vocalizations that are the most common means of social communication among most dolphin species, but orcas whistle only occasionally. They have a different type of vocalization that makes up most of their social communication: the discrete call. Discrete calls are short (one- to two-second) tonal vocalizations. Each call has a distinct structure that is stereotyped, meaning it is mostly unchanging over time—at least so far as researchers have been able to observe. One call may sound like a single ascending note, another may be a relatively flat tone before downsweeping at the end, while a third may bounce across different frequencies, rapidly modulating up and down. These unique sounds can also be visualized via a spectrogram—a graphical representation showing a sound's frequency, duration, and amplitude.

In the Pacific Northwest, every call, like every whale, has been given an alphanumeric designation. "S" calls belong to the Southern Residents, "N" calls to the Northern Residents, and "T" calls to the transients. Among the Southern Residents, common call types vary by pod, meaning it's possible to tell which pods are in the area just by their vocalizations.

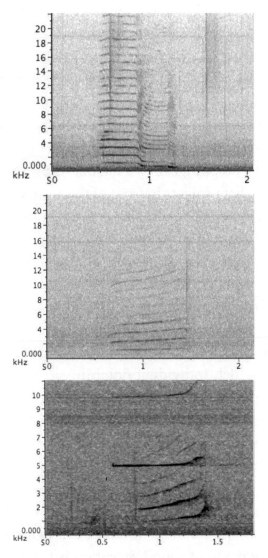

Spectrograms are a way to visualize sounds. Here, time (in seconds) is on the x-axis and frequency (in kHz) is on the y-axis, with the amplitude of the sound visible as the intensity of the color in the image. Each killer whale discrete call has a unique sound and spectrogram. Pictured here are the most common call types of (a) J-Pod (S1), (b) K-Pod (S16), and (c) L-Pod (S19). Generated in Raven Pro 1.2.1 from recordings made by the author.

Whales and dolphins do not produce sounds like humans do, by passing air through the larynx and over vocal cords. The mechanisms of cetacean sound production are not well understood, but odontocetes produce sounds using nasal air sacs in their heads. As air is transferred between the narrow passageways between the sacs, the vibration creates noise. To gain a better idea of how this mechanism works, imagine squeezing the mouth of an air-filled balloon to create various squeaks and screeches. The fatty tissue in the forehead of a dolphin, called a melon, transmits this sound into a beam that projects in front of the animal. Using these physiological features, dolphins can control with great precision the qualities of the sounds they make, produce multiple sounds simultaneously, and control whether the sound is projected in a broad band or a narrow band in front of them. Due to the way sounds are produced, dolphins don't have unique voices like humans do. Rather, they can manipulate the pitch of their calls, so it is not possible to determine which individual dolphin you are hearing just by the tone of the call. Not all sounds that cetaceans make are produced vocally. Surface percussive behaviors, such as breaching and tail slapping, generate underwater noise and may play a role in acoustic communication.

In addition to their unique method of sound production, cetaceans receive sound far differently than terrestrial mammals. While some sound is received through the ear canal, the primary mode of sound reception is through the lower jaw, where bones and tissues conduct sound directly to the inner ear. In his early research at the Vancouver Aquarium, Paul Spong experienced this firsthand when Skana, a captive orca, would place her lower jaw near a new sound source to examine it more closely. Whether it was music from an underwater speaker or Spong creating a high-pitched ring by running his finger over the rim of a wine glass, it was Skana's lower jaw, not her ear, that she placed against the object to hear it better. It's also likely that killer whales can passively listen to the sounds and echoes from other whales' vocalizations. For instance, a whole family of orcas traveling close together may be able to construct an acoustic map from the echolocation click trains emitted by just one of their members.

ALEXIS'S MORNING AT SALMON BANK

It's difficult to imagine living in a world dominated by sound. Using what researchers know, combined with a healthy dose of my imagination, let's indulge in what it might be like by spending a morning as L12 Alexis, a female in L-Pod and matriarch of the L12 subgroup.

With a slight flick of her flukes, Alexis glided almost effortlessly over Salmon Bank as she had done countless times in her seventy-plus years of life. Her family was spread out over several miles off the south end of San Juan Island, each individually hunting fish, except for the calf Cousteau, who closely followed the lead of her mother, Calypso. They had been foraging for the better part of the night, but now the zooplankton were descending back into the depths and the top layer of water was brightening— dawn was coming. At the moment, all was silent. Alexis didn't have to echolocate to find her way across the underwater shelf; all her years of hunting over the bank had given her a perfect mental map of every nuance of the seafloor.

A sudden popping sound from below caught her attention. She turned her head to look below and sent out a short click train directed at the source of the sound. The echo came back quickly off the bottom—it was shallow here—but it was the subtler reflection she was interested in. The gradations reverberated back through her lower jaw and directly to her inner ear. In an instant, Alexis matched the sound pattern to one she had experienced many times before: it was a large rockfish, but she wasn't interested. Suddenly one of the other orcas called out. It was an S2iii, the contact call of Alexis's family group. Every time a member of the pod made this call, it had the same overall structure, but it was the variations in frequency and duration that held all the key information. Alexis recognized Mega's signature within the call, and from the single vocalization she learned that he had decided to move further offshore toward several purse seiners fishing nearby. After a brief pause, two more calls from the north responded, indicating that Mega's sisters Matia and Calypso were going to meet him there. Alexis called back, making sure to broaden the course of her call so all of her family members could hear it despite being in different directions; she was coming too.

Turning toward Mega, Alexis started swimming. The tide had been slack, but she noted the slight push of the incoming flood tide beginning. Experience told her that the fish they were after—Chinook—would be on the move before too long, making the next leg of their migration to the fresh water river where they would get out of reach of the orcas. But she also knew that today the tides would be mild, and some of the fish that made it in from the ocean

L12 Alexis at Salmon Bank, where she often foraged with her family, the L12 subgroup.

yesterday would prefer to wait for a larger tide to push them north. Alexis and her family should find enough to eat here today. Long before she reached Mega, Alexis heard the familiar chugging hum of the fishing fleet. Each vessel had a distinct signature, like each member of Alexis's pod, but for some reason the boats thought it necessary to continuously broadcast their call when moving. Some of the human-caused noises annoyed Alexis, but not this sound. They had been present as fishing companions for so much of her life that they were an accepted part of her environment. She could tell that they were moving slower than normal. Alexis spyhopped to confirm what she suspected: there was a dense layer of fog over the water's surface, causing the fishermen to be sluggish in their work. With the fog, as at night, they had difficulty seeing what was going on. How strange, Alexis thought, to be so dependent on vision.

She focused on the task at hand. As Mega had indicated in his call, these seiners had successfully found a concentration of salmon. Having caught only a few good fish during the night, Alexis was hungry. She sensed the movement in the water off to her left and knew instinctively by the size of the disturbance that it was Mega, her grandson. The nearest seiner was in the middle of placing a set, and the two whales swam side-by-side toward the open arc of the net that spanned the distance from the seiner to the skiff. Both ends of the net were beginning to herd the salmon, and Alexis and Mega each headed toward opposite ends in a tactic her family had used for decades. The two whales acted as an additional barrier, effectively closing the circle as they swam toward the net, pushing more salmon into it. Once they reached the outer ends of the curved net, Alexis turned toward the middle where the salmon were densest, easily picking out one of the larger fish. She returned to Mega, and they tore the Chinook in two, sharing it.

To an outside observer, their fishing method might look like a graceful dance, each whale twisting and turning in perfect synchrony with the other. It was impossible to see much of anything even just a few dozen feet below the surface of the water; instead, Alexis and Mega were guided by the sound of the skiff and the seiner, the swishing of the salmon as they tried to flee from the barrier presented by the net, and the rush of moving water created by the other whale. This was a choreography the two whales had enacted together many times, and Alexis recalled how she used to engage in the exact same foraging tactic with her own son, Okum, before he passed away. Mega had learned the method from his mother, Alexis's daughter Squirty, and undoubtedly the same approach would be used by future descendants as long as there were orcas, and salmon, and purse seiners.

Mega and Alexis shared another fish, then Calypso approached with Cousteau in her slipstream (the wake created by the larger mother that helps to pull her calf along). Cousteau, enthusiastic at seeing her extended family after an absence of several hours, bolted ahead of her mother to where the two larger whales were feeding. The young calf was ready to head into the curve of the net to start chasing the fish like her uncle had done, when Alexis cut her off with a sharp rake of her teeth against the little whale's back. This brought Cousteau to a screeching halt, and with a short vocalization Alexis communicated to the young whale that this was not a time for play. The nets presented a danger worth being mindful of. The calf darted back to the safety of her mother's side, disappointed at being left out of the fun of playing in the net, but Alexis knew better. She had twin scars on her dorsal fin—a result of her own carelessness around nets as a calf. Luckily her mother had been quick enough to skillfully disentangle Alexis from the ropes before any worse damage had been done than the two minor wounds to her fin and the injury to her ego. Alexis didn't want Cousteau facing such risks.

The family caught one more fish before the circle of the net began to close and it became too narrow for the whales to snatch the surrounded salmon. Another S2iii vocalization reached them over the whirring of the boat engines. It was Matia, telling them she had just finished feeding as well. The four orcas left the seiner to reel in the rest of their haul, heading in Matia's direction. While swimming, Cousteau slipped below Calypso and nursed for a few minutes. Now that the whales all had full bellies, there was time to play. After being apart for most of the night, the orcas rejoiced in each other's company, each whale swimming alongside the others and stroking their bodies with pec fins and tail flukes. With a glance to Alexis for approval, the calf Cousteau realized that now was a good time to let out some of her pent-up energy. She squealed a loud version of the S2iii call. It was a good attempt but failed to match the structure of the intended call, much like a human baby babbling in an attempt to form words. Cousteau's aunt, Matia, was amused by the vocalization and emitted a call saying as much. Calypso playfully head-butted her sister Matia for teasing her calf, and soon enough all the whales were off, chasing each other in circles.

When Mega headed up for a breath, he lifted his flukes and exuberantly crashed them down onto the surface of the water, creating a *slap* that echoed across the banks. Cousteau, a fraction of her uncle's size, swam up beside him and threw her own tail slap, which resulted in a quiet clap of sound compared to the thunder of Mega's tail hitting the surface. Mega rolled over under his niece and surfaced again, executing a slower tail slap to show her the proper technique. Cousteau followed suit, achieving better form on this attempt but still barely creating a ripple of noise with her small tail. Alexis had been watching from below. She gave several strong pumps of her tail flukes and launched right past Mega and Cousteau through the surface of the water. For less than a second she was airborne, experiencing the strange sensation of gravity and the intense brightness of daylight, before she came crashing down with a booming splash far louder than the tail slaps by either Cousteau or Mega. The other whales erupted with laughter.

Times like this were important to the whales. Not only were they having fun, but Cousteau was learning from the adults, and all of the orcas were reaffirming their familial bonds. Being lifelong companions, it was vital to a family group to spend time enjoying each other's company and strengthen the trust that was essential to their survival. Even though they were always within acoustic range of one another, it was a pleasure to spend time close enough to see each other. Orcas are tactile creatures, and the caresses and touches they exchanged were an indulgence. After a few more minutes of play, Alexis surfaced for a breath while the rest of her family continued in their cavorting. Tired from her exertion, she decided to hang at the surface and catch a quiet nap while she logged, floating at the surface to rest. Still partially conscious, she kept an ear tuned in to her family's squeals and whistles as they swam a little further away.

Distantly, Alexis thought she heard another vocalization. She listened more closely and heard it again. It was an S1, a J-Pod call, but who was it? It was still too faint to tell. After a moment, the call sounded again, far louder than before. It was unmistakably Granny, calling out specifically to Alexis to tell her she was coming. Alexis took another breath, alert now, and blinked a moment in confusion. Taking in her surroundings, once again she heard only her family members nearby. Had it just been a dream? It reminded her of something, a conversation she had with Granny more than a week before. Their extended family, J-Pod, was due to return soon, and as Alexis tuned in again to the water temperature, the strength of the tides, and the particular current flowing this morning, she recalled that today was the day.

Alexis turned toward her family group and called out with an S6, a call type they shared with J-Pod, to tell them the news that the extended family would be returning to inland waters today. The playing immediately ceased, but a sense of excitement remained as Calypso responded with another S6, filled with a note of questioning. In response, Alexis sent out an affirmative S6—yes, she had decided, they would all go to meet the incoming J-Pod. Picking up speed, the whales took up their ranks on either side of Alexis. Mega, being the oldest, was at her side. Next to him were his sisters, Matia and Calypso, with the calf Cousteau between the two adult females. The family group headed west, each whale silent as they thought of meeting up with the J-Pod orcas.

WHAT ARE THEY SAYING?

It's compelling to imagine how Alexis and her family might communicate, but do we really know what orcas are saying? In the late 1970s Canadian scientist John Ford started studying the discrete call vocalizations made by wild killer whales in British Columbia and Washington. While orcas can hear from 0.5 kHz to 125 kHz (compared with 0.2 kHz to 17 kHz for humans), most of the energy of the calls occurs within 1 and 6 kHz, well within the range of human hearing. As a result, each call type sounds unique to the human ear and has a signature spectrogram. Ford

cataloged all the vocalizations he heard, eventually identifying distinct call types made by different populations
of whales. The Northern Residents made more than forty
different discrete call types and subtypes, the transients
eight calls, and the Southern Residents twenty-six calls.
Future research showed that different killer whale social
groups had different dialects. Some calls were shared at a
clan level, others at a pod level, and still others only at a
matrilineal level.

For example, the Southern Residents, which make up
a single acoustic clan, don't share any calls with the three
acoustic clans of the Northern Resident community or
with transients. At the clan level, call types have proven
to be static over the course of the fifty years for which we
have recordings. The repertoire of each individual pod is
mostly maintained, with the dominant call types for each
pod staying the same. For example, S1 makes up about 25
percent of the vocalizations of J-Pod, so an experienced listener would be able to identify a J-Pod recording regardless
of if it was from 1975 or 2015. Neither J-, K-, nor L-Pod
make all Southern Resident call types, but rather each pod
has a repertoire made up of a subset of those calls. Some
calls are shared across two or three pods, while others are
unique to a single pod. It turns out that calls are markers of
different cultural groups, varying across the different levels of orca social structure. Acoustic similarities between
groups of killer whales have also proven to be an accurate
measure of relatedness. The greater number of calls two
pods share, the more genetically similar they are. Acoustic
similarity reflects shared ancestry more than social associations. Those pods that spend the most time together

do not necessarily sound the most similar. While related groups of orcas share more call types, the types of calls they make are not genetically determined. Instead, call types are learned.

While dialects are common in birds, they are relatively rare in mammals, with nearly all mammalian vocal differences being due to geographic isolation and/or genetic variance. Only three types of mammals have demonstrated vocal dialects among groups that have overlapping ranges: primates (including humans), bats, and cetaceans. In these situations, where dialects are learned rather than genetic, differences in vocalizations act as a type of social adaptation. For instance, dialect differences may lead to assortative mating, wherein you only mate with those who sound different from you in order to make sure you are avoiding potential inbreeding. Ford suggested that the reason resident killer whales have large repertoires of call types is partially to serve as a badge of group affiliation. If you share all your call types with your immediate family and some call types with your more distantly related family, this may help whales both strengthen group bonds with family members and identify ideal mates that are not related.

From observations of killer whale calves born in captivity, a bit more is known about the vocal learning that occurs in young whales. Calves begin to produce sounds within a few days of birth, although these noises are not representations of adult vocalizations but rather various screams and tones akin to babbling in human babies. Young calves selectively learn the calls that their mother makes, even when they regularly interact with other whales with completely different repertoires of calls. This coincides with what we know about wild whales; a calf will ultimately have the same repertoire of calls as its mother and her matriline, even though calves are raised in an environment that comes into contact with other groups that make different vocalizations. In addition, researchers from SeaWorld have reported that calves start making calls similar to adult discrete calls around the age of two months; their call repertoire expands thereafter until they reach adolescence, when they have acquired their full range of calls.

During the summer of 2003, my first year as an intern at The Whale Museum, another student and I spent long hours at Lime Kiln listening to recordings, learning the nuances of each orca call. After long periods in silence, sitting side by side in a small shack next to the lighthouse known as the "acoustics shed," we pulled off our headphones and puzzled over a call. "Is this an S7 or an S12?" We checked our battered paper copy of Ford's call catalogue, the exclusive reference, since it included spectrograms for every call type he identified. After carefully comparing spectrograms, we noted the differences or similarities to the call in question. We played clips of call types saved from previous recordings and likened them to the unidentified call. Through this practice and repetition I learned to differentiate the killer whale calls from one another. An S7 call has a gap between the second and third parts, and the third note is monotone. For an S12, the second and third parts of the call run into one another, and the third part has a distinct down-and-up component that appears in the spectrogram like the letter U.

As I puzzled over and learned the different call types, I pondered how the whales used the calls. Did each call function like a letter within our alphabet or a word within our language? I didn't have to listen to too many recordings of Southern Resident killer whales to begin to doubt this is the case. They don't often make complex chains of different call types as one might expect if each call were a component of communicating a longer complex message; instead, the orcas often make one or two call types in long chains, repeated by a single whale or by different whales calling and answering to one another. Rich Osborne and his research partner, Rus Hoelzel, reported that about 80 percent of their recordings made between 1977 and 1981 were repetitive sequences, defined as one call being repeated five or more times in succession.

Another indicator that discrete calls probably don't function as units of information is the ability for different groups of whales to understand each other regardless of the number of shared call types. For example, all three pods in the Southern Resident community regularly interact despite having only a handful of shared call types among the pods. It is even more striking in the Northern Residents, where there are three different acoustic clans that don't share *any* call types, yet they regularly socially interact. This suggests that having a different repertoire of calls is probably not akin to speaking a different human language; rather, the unique call types are perhaps a badge of group identity, and more detailed information on things like emotional state and current behavior are likely contained within those calls in a manner that is more universally understandable by all killer whales.

Ford and other early killer whale acoustic researchers posed their own sets of questions about orca call use. Were certain calls used only in specific contexts, like when traveling or feeding? Could call types be associated with a particular behavior? The answers further proved just how complex killer whale call use is. While some calls occur more or less often depending on behavioral state of the group, nearly all call types occur across all behavioral categories. For instance, while you might hear J-Pod make the S4 call most often when traveling, the call also occurs regularly during socializing, meaning it is not specific to a particular type of behavior. Further evidence of call types not being context-dependent comes from recordings of captive whales. Despite being in environmentally and socially static situations, captive orcas produce the entire call repertoire of their wild family group. In the late 1970s and early 1980s, researchers Alexandra Morton and David Bain conducted separate acoustic studies of different captive whales. They failed to find any 1:1 correlation between a call type and a specific behavior. This means that the whales aren't using a particular call to mean "breach" or "fish." In Morton's study, however, she compared the relative frequency of call use to different activity states and found something interesting. Three of the call types she looked at were associated with three different moods: one occurred far more often when the whales were calm, one when the whales were playing, and another when the whales were in a stressful situation. This finding suggests that vocalizations might be more related to the emotional state of the animal rather than the specific behavior they are engaged in.

If each call type indicates something about the emotions of the animal, perhaps more complex messages are contained within the minor variations within the call. To humans, each S1 sounds fairly similar, although if you look closely at the spectrograms for these calls there are subtle differences in each call's components. Dolphin brains are wired to process acoustic information in a way that we humans can only imagine. To the whales, the call categories as defined by researchers may not be what is most meaningful at all. Perhaps the subtle variations within each call—those beyond our perception to easily understand—contain the important information.

THE ACOUSTIC MYSTERY CONTINUES

In 2010, I was contacted by Andy Foote and Nicola Rehn, two killer whale researchers in Europe who were conducting an acoustics study. They had a set of sixty-nine orca discrete calls that they asked me and eight others familiar with listening to killer whales to categorize. I was eager to do it. It was like getting a glimpse into the task that was before John Ford in the 1970s—the thrill of taking a set of recordings and sorting what you've heard into call types. I sat in front of the computer playing and replaying vocalizations over my headphones. Some of the calls I thought I recognized as being Southern Resident calls, but most were clearly from a different population of orcas. On a piece of paper, I drew out a little picture for each sound, imagining what the spectrogram might look like. After doing this for each clip, I lumped the ones together that had a similar structure: eight calls were ascending, fourteen calls were descending, and so on. I went through each group again, splitting out each broad category into what sounded to me like specific call types.

When I was done, I had sorted the sixty-nine vocalizations into eighteen different call types, with two to sixteen samples of each one. When I sent Foote and Rehn my results, I made a note on the call type with sixteen samples. "These sixteen call types show some variation," I wrote, "but seem closely associated." To me, they sounded like the S10 call I was familiar with from the Southern Residents. This call, made by all three pods but especially common during superpods, has a lot of tonal variation, going up and down and lasting several seconds. On many occasions I've heard listeners comment that this call sounds like human laughter. Foote and Rehn were excited by what I had written. They, too, thought these sixteen vocalizations were the same call type, but what was amazing is that they didn't come from the same population of killer whales. Instead, within that set of calls were recordings from Southern Residents, all three acoustic clans of Northern Residents, Kamchatka Residents, offshores, and Alaskan transients. "We call this the 'excitement' or V4 call type," Foote explained. "It seems this call is not learned like the rest of the repertoire, but is innate and universal—a bit like laughing or crying in humans!" I couldn't believe it. Perhaps the people who associated the S10 call with laughter were right!

Foote and Rehn published their results six months later. The fact that nine listeners blind to the source of the calls came to the same conclusion helped strengthen their case that this was, in fact, a universal call type across different killer whale ecotypes. What was particularly interesting was that the call occurred during excited states regardless of which group of whales was making the call. For example, the resident recordings were made in multipod aggregations and the transient recordings were made shortly after a prey kill. Their findings further bolstered the idea that specific call types are related to emotional states rather than being linked to specific behaviors.

The way killer whales communicate is so foreign to us that our glimpses into what their calls might mean are few and far between, but on rare occasions we might have an observation that gives us further insight. One of my most memorable acoustic encounters with orcas occurred in August of 2004. It was late evening and the L12s, a sub-group of L-Pod, were heading south in Haro Strait toward the Lime Kiln Lighthouse. Sitting low on the rocks, I noted a few other whale-watchers above me. The whales were blowing in the distance, and it looked like they were about a half-mile offshore. Suddenly we saw a blow much nearer: one animal had broken off from the rest of the group and was swimming directly for shore. To our delight, right when it got to the patch of kelp in front of the people gathered on the rocks, the creature stopped and turned to face us. With just its forehead and part of its dorsal fin above the surface, the whale vocalized into the air, repeating the same sound again and again. I immediately recognized the call as S2iii—coincidentally the same call type I had identified for Rich Osborne as a blue-haired teenager a couple years earlier.

The whale vocalized for about two minutes, repeating the same call every few seconds. I had heard whales make a single brief vocalization above the surface before, but never like this. Everyone watching had a strong reaction to witnessing this whale calling at the surface. People asked questions out loud: Are you okay? Who are you? And, most strongly in my mind, what are you trying to tell us? As the rest of the pod caught up to our gathering on the shoreline, this whale took one last breath and turned away, heading out to rejoin its family group and continue traveling south. Although the entire experience lasted only a couple of minutes, it is one of my most vivid whale encounters ever. It's the only time while watching whales that I ever strongly felt that *they* were trying to communicate with *us*. My frustration about not being able to understand what the orca communicated was immense. Later I described the incident to Osborne. After a thoughtful look, he simply said: "Hang onto observations like that. They could be like Rosetta Stones."

This whale was either a young male or a female, which was apparent because it had a shorter dorsal fin. Because of backlight and the fact that the whale surfaced with kelp draped over its fin, it was impossible to make out the saddle patch. As a result, its identity remained a mystery. Four years later, dedicated whale-watcher Jeanne Hyde, who had also witnessed what we took to calling "the above-water

L85 Mystery with a dorsal fin covered in bull kelp as he vocalized into the air for several minutes at Lime Kiln.

vocalization incident," went back to her video clip of the encounter and took a still image of the eyepatch, which was visible on one surfacing. Comparing this to other eyepatch photos she had taken in the intervening years, Hyde was able to confirm that the whale was L85, also known, appropriately, as Mystery. At the time of the incident, he was thirteen years old.

The only knowledge I've gained about the mysterious incident since it happened is that I am more familiar with the S2iii call. By far the most common vocalization used by the L12s, it's probably the contact call for that group of whales. Each pod has a single call type that occurs far more often than all the others, and this call perhaps functions to keep different pod members in contact with one another when they are out of visual range. In the murky waters of the Pacific Northwest, that is most of the time. For J-Pod, it's the S1 call. For K-Pod, the S16. Most of L-Pod uses the S19 call, but as the L12 subgroup spends more time apart from the rest of L-Pod. They have started to acoustically diverge, and use the S2iii call most often. The only thing I can conclude from the whole experience of hearing Mystery vocalize at the surface, at the only place that evening where humans were gathered onshore, is that he was indeed trying to make contact.

LEARN MORE

For more on killer whale acoustics, see various articles by Volker Deecke, Andrew Foote, John Ford, Marla Holt, and Nicola Rehn. To hear killer whale acoustic samples and listen for them live, visit Orcasound (the Salish Sea Hydrophone network) at www.orcasound.net.

CHAPTER 4

Personality, Emotions, and Culture

J27 BLACKBERRY WAS THE FIRST orca I symbolically adopted, before I had ever even seen the Southern Residents in person. I remember my excitement at twelve years old as I opened an envelope from The Whale Museum and pulled out the certificate with Blackberry's picture on it, which I pinned up on the bulletin board in my bedroom. The photograph caught my eye, but it was Blackberry's personal history that especially piqued my interest. He was born in 1991 to J11 Blossom. He had a younger sister, J31 Tsuchi, whose name comes from the Japanese word for "beaked whale." He was often seen playing with J26 Mike, another young whale in his pod. With just a few simple facts and stories, the whale pictured on my certificate became an individual with a unique personality and characteristics. I began to realize just how well known this population of whales was.

It took a while before I was able to pick Blackberry out of the crowd of Southern Residents. When I first saw him, his fin and saddle patch looked like every other female and juvenile's fin at about two to three feet tall. The adult males were the easiest place to start learning individual identifications, in part because their larger fins made them easier to pick out but also because there weren't many of them at the time. Unfortunately, in the late 1990s there had just been a die-off of many of the adult males, and the summer I began learning the family groups there were only four adult males in the entire population of Southern Residents. They were J1 Ruffles, the easiest whale of all to identify with his wavy dorsal fin; L41 Mega, the only one with a notch in the middle of his fin; L57 Faith, whose dorsal fin had a distinct lean to one side; and L58 Sparky, who had the most generic fin of the bunch with no obvious distinguishing features. I slowly learned how to identify more of the whales, and by a few years later, Blackberry and his friend Mike were the first two young males I got to see "sprout" as they grew to have the tall dorsal fins of adult males. Mike's fin began to lean to one side much like L57's, and Blackberry's developed a slight twist that made the angle of his fin distinct. Even though there were a lot of whales I didn't know, and most of the time I felt like a distant observer, Blackberry was one of the first whales I watched grow up. As a result, my connection to him felt stronger.

There was a big change in Blackberry's life in 2003, when his mom Blossom gave birth to Blackberry's younger brother, J39 Mako. Blossom, Blackberry, and Tsuchi had been a pretty tight-knit family group. Sometimes with the arrival of a younger sibling, an older whale may start roaming a little further and not spending as much time by its mother's side. I wondered if this might happen for Blackberry, but the new calf fit right in and the close family group of three became a close family group of four. In fact, when Mako was old enough to start becoming a little more independent, he took to his big brother as the whale to hang out with. Instead of being babysat by his older sister Tsuchi, Mako often swam right by his big brother Blackberry. One evening in June, late in the day as I was at the top of Lime Kiln Lighthouse leading a tour, the whales came by. I recognized Blackberry heading toward us from the north, and there was another little whale with him. The group was close enough to shore that we looked straight down from the top of the lighthouse to see them swimming underwater. I watched the large, dark shape of Blackberry slowly come to the surface. Just in front of him was the smaller whale, whom I now was able to identify as Mako. Mako rocketed to the surface near his brother, nearly clearing the water.

A few weeks later, I witnessed another interesting event involving Blackberry, Tsuchi, and Mako. They were toward the end of a group of J-Pod whales, slowly making their way north one morning. While most of the whales were in travel mode, these three siblings were lunging at the surface and doing lots of circling, the sort of behavior that's more commonly associated with foraging. Suddenly I saw something pop up in the middle of their circle—were they pursuing a salmon? I snapped a few photos, and it was only upon going back and looking at them later that I realized it wasn't a salmon they were playing with but an adult harbor porpoise.

PORPOISE HARASSMENT

Resident whales are considered to be exclusively fish eaters, leaving the marine mammal side of the food chain to their transient killer whale cousins. This was the first time I had

seen Southern Residents in active pursuit of a marine mammal, but as I did more research into the topic, I learned that
this behavior is not altogether uncommon. Although there
has never been confirmation of a resident orca consuming a marine mammal, they are periodically seen chasing,
playing with, pushing, and sometimes killing porpoises.
They harass the porpoises, sometimes to the point of death,
and while they will even spyhop with the porpoise in their
mouth, there is no evidence they are actually eating them.
The first documented incident of the Southern Residents
engaging in this bizarre behavior was in 1976, the same year
in-depth research on them began.

From 2003 to 2016, I documented the occurrence of
more than thirty incidents of Southern Residents actively
engaging with porpoises in this manner. Whale researcher
Robin Baird reports that the occurrence of porpoise kills
was relatively uncommon for several decades until the
mid-2000s, when the number of incidents spiked. It has
remained more common since then, perhaps due to the
recovery of the regional harbor porpoise population,
making encounters between the two species more frequent. The majority of the incidents involved adult or
neonate harbor porpoise, while the rest of the encounters involved Dall's porpoise. L-Pod whales are the most
likely to engage in this behavior, although all three pods
have been seen doing it; even legendary J2 Granny was
seen pushing around a harbor porpoise with L87 Onyx!
The J11s, J14s, L11s, and L54s have been seen engaging in
this behavior numerous times. Often multiple whales are
involved, and in the majority of encounters, at least one of
them is a juvenile.

J40 Suttles balances a harbor porpoise on her head. Photo by Sara Hysong-Shimazu.

The reason for this behavior remains obscure, although several theories can be ruled out. The orcas aren't killing the animals for food—often they abandon the carcasses, and no prey studies have ever found evidence of marine mammals being a part of a resident killer whale's diet. Eliminating potential competition for food is also unlikely, since harbor porpoises commonly feed on herring and sculpin, while killer whales focus on larger species like salmon. Predators sometimes kill animals and don't eat them when they are training their young, which may seem like a plausible theory since juvenile orcas are often involved in these porpoise events. But again, the question is raised of what they would be training them for if porpoises aren't a regular prey item. In addition, while many of the porpoise carcasses show puncture wounds from orca teeth as well as rake marks, necropsies on a few of the animals have shown that they have likely died of trauma rather than any obvious internal injuries or exterior lesions that are more likely to be associated with a predation event.

So why would whales like Blackberry, Tsuchi, and young Mako be seen playing with a harbor porpoise? The most likely answer seems to be the key word in that last sentence: play. Killer whales are the undisputed top predator in the ocean, and while different populations specialize on different prey items (e.g., fish, porpoises, sea lions, sharks, rays, or other whales), every member of the species is equipped with the same adaptations that make them efficient hunters. Perhaps some of this killer instinct takes over, or perhaps it is just a game for the orcas to engage in: to play cat-and-mouse with a porpoise. A lot of questions came up for me as I saw Mako learning to torment a porpoise from his older siblings, but like so many questions about cetaceans, these answers for now remain shrouded in mystery. These bizarre incidents of fish-eating whales killing porpoises just adds to the intrigue of the Southern Resident population. They remind us that even though the resident killer whales sometimes are characterized as being "friendlier" or "nicer" than the transients, they too are efficient predators.

FUN AND GAMES

Porpoise harassment isn't the only behavior resident killer whales engage in seemingly just for fun. Northern Resident orcas are known for their rubbing beaches, steep beaches in shallow waters where they go to rub themselves on smooth, round stones. This is a highly ritualized behavior for them:

it only occurs in specific areas, and they show great excitement when approaching beaches to rub. With fewer similar beaches in their core range, the Southern Residents aren't known to rub on rocks.

What Southern Residents do seem to like is the tactile sensation of bull kelp—a thick, brown seaweed that grows in long strands in shallow coastal areas. Bull kelp is anchored to the seafloor via a holdfast. A long, thick stipe (the main stem of the kelp) extends from the bottom to the water's surface, held aloft by a float. Attached to the float are as many as thirty to sixty wide, flat strands that drift in the current, creating the dense canopy in what can be a lush underwater kelp forest. An annual plant, at the end of the season bull kelp breaks free from its holdfast, so in addition to washing up on beaches, mats of kelp can be found free-floating in the water. "Kelping" is the act of swimming through a kelp bed. Sometimes a whale will stop and roll around in the kelp, surfacing with it completely draped over its body. One can imagine this might be the equivalent of a whale massage. The orcas also like to play with the kelp. They surface with it wrapped over their dorsal fins, lift their tails into the air with kelp draped over it, push it around with their rostrums, or even carry it around in their mouths.

The behavior known as kelping has long persisted among Southern Residents, but there are other unusual behaviors that are passing fads among the whales. They're discovered, gain popularity as other whales learn to do them, and then fade perhaps as the whales become bored with them. One such example occurred in 1987, when a K-Pod female started surfacing with dead salmon balanced on her head, apparently for fun. Over the coming weeks many other whales started

Southern Resident killer whales seem to enjoy swimming through kelp beds, draping the fronds over their fins, carrying pieces in their mouths, or stopping to roll around in it.

mimicking this same behavior, until the novelty apparently wore off about six weeks later. The behavior was observed a couple of times the following year, and then not again. This behavior doesn't have any benefits to the whales in terms of survival; it's simply a cultural fad that the whales thought was fun for a little while. Other examples of similar fads over the years include pulling buoyant kelp bulbs underwater and letting go so they rocket up into the air, and spyhopping with objects such as driftwood held on the pectoral fins.

Many of us spend a lot of time watching killer whales, but do they ever spend time watching us? For the most part, the Southern Residents seem to go about their business of traveling, foraging, resting, and socializing regardless of whether we are watching or not, but occasionally, like the encounter with L85 Mystery at Lime Kiln, there are rare moments where the curiosity between orcas and humans seems mutual. On a rainy May day in 2005, as part of my training, I tagged along aboard the *Western Prince* to shadow a naturalist in action. It would be my first summer working as a naturalist on a whale-watching boat. The *Western Prince* was less than half full on this particular Tuesday when we headed south from Friday Harbor, out through Cattle Pass, and met up with some J-Pod whales near False Bay. We were following the J17 matriline, then made up of J17 Princess Angeline and her two young daughters, J28 Polaris and J35 Tahlequah. Princess Angeline, with her unusual saddle patches, and Polaris, with a tear in her fin, were easy to identify. Tahlequah was an average-sized whale at seven years old. Her nicely curved dorsal fin and solid saddle patches made her more difficult to identify at a glance, so I often knew her simply as "the whale traveling with Princess Angeline and Polaris." It was on this day, however, that she would become an individual to me.

The boat parked so that the whales would pass off the stern about 150 yards away (the guidelines at the time were for vessels to stay one hundred yards away). I stood under the awning a few steps from the back of the boat when Polaris and Princess Angeline surfaced right where we hoped they would. I waited for Tahlequah to pop up when one of the passengers standing in front of me out in the rain pointed down, shouting, "Look at this!" I rushed forward to kneel on the seats at the back of the boat and looked down to see the dark shadow of Tahlequah approaching the stern. As she neared us, she rolled onto her left side, using her right eye to look up at those of us leaning over the railing. After she passed, seemingly without moving a fin, she turned and glided back toward her mom and sister. The next time Tahlequah surfaced, she

Occasionally, the curiosity is mutual. A young killer whale checks out our boat.

was back alongside Princess Angeline; no part of her body broke through the water as she took a little detour to check us out. There was no mistaking that was her intention. When you look at a photograph of a whale, you get a static image of the animal, often just the dorsal fin, with no indication of the individuality that lies beneath. As you observe these animals in person, however, and get to know them, you glimpse the moods and personality traits of each whale: laziness, playfulness, or in Tahlequah's case, curiosity.

I came to know Tahlequah not only as curious but playful as well. During a few months from 2009–10, the J17s doubled in size as Princess Angeline and both her daughters had calves. With the addition of J44 Moby, J46 Star, and J47 Notch, the J17s became known as the "nursery group" for the next several years. Whenever we saw them at least one of the calves was usually being rambunctious, more often than not goosing one of the other kids or the adult whale in charge to play along with them. One day in 2010, when Tahlequah's son Notch was just a few months old, I was again aboard the *Western Prince*. Tahlequah, whom I practically thought of as still a kid herself since she was a young mom at the age of twelve, seemed to be showing her young son how to play. Mom would spyhop, calf would spyhop, then the two would spyhop together. The same thing was repeated with tail slapping. Notch was certainly mimicking everything Tahlequah was doing. She, even more than the other moms, always seemed up for play.

EMOTION AND CULTURE IN NONHUMANS

Over the past half century, studies of animal behavior have broken down the barriers of what we thought made us uniquely human. We've learned that chimpanzees fashion sticks to fish termites out of their nests (tool use). Prairie dogs make different alarm calls to alert each other to the presence of different predators (language). Monkeys won't accept food if it means another monkey will receive an electric shock (morality). Dolphins recognize themselves in mirrors (self-recognition). Horses can be introverts or extroverts (personality). Elephants grieve the death of family members (emotion).

Conservationist Carl Safina has written extensively about our perceptions (and misconceptions) of emotions in other animals. Killer whales are one of the three main species he focuses on to make his argument for the complex emotional lives of other species. Since we all evolved from a common ancestor, humans are more similar to other species than we might admit. The capacities we have as humans didn't start with us; rather, they occur widely in other animals that share similar brain structures. Animals likely have thoughts and feelings more similar to humans than many people may realize. Those of us with pets know quite well how individual animals (mammals, birds, and even reptiles or fish) have different personalities, the ability to reason, and a sense of humor. Why are we so reluctant to attribute the same characteristics to wild animals? Perhaps because it is a very gray area; it is all too easy to prescribe human emotions to animals (i.e., anthropomorphizing) when something very different may in fact be going on. As a result, many people go to the other extreme and argue that it isn't possible to conclude animals experience emotions at all.

The more we learn about our fellow animals, the more we realize we're not so different from them after all. Like humans, animals such as orcas that live in long-term, stable social groups exhibit behaviors that are unique to the group in which they live and socialize. Such socially learned behaviors are the basis by which we understand culture; among biologists, "culture" is defined as shared, persistent, socially learned behavior that is not explained by genetic or environmental factors. It is a relatively new idea to attribute culture to any species other than humans, and yet among cetaceans we see undeniable hallmarks of culture. Perhaps one of the best examples comes from the two distinctive ecotypes: the fishing-eating resident killer whales and the mammal-eating transient killer whales. These different lineages inhabit the same region but live very different lives. We can liken this phenomenon to human cultures that developed into a pastoral versus agrarian way of life.

Communities form around those who live the same way, leading to unique traditions within those communities.

Language, art, religion, politics, morals, games, and gender roles can culturally separate one human community from another based on the shared behaviors within but not between the communities. While killer whales may not engage in art, religion, or politics, they do exhibit community-specific behaviors that are learned within a specific social context. This argument, made by biologists Luke Rendell and Hal Whitehead, offers various examples of culture among different cetacean species, particularly four of the most well-studied groups: bottlenose dolphins, humpback whales, orcas, and sperm whales. The response to the work of these two biologists demonstrated how the idea of culture in other species remains a controversial one. Despite the critiques of the evidence Rendell and Whitehead published, it has become more accepted to explore culture as an explanation for behaviors observed in cetaceans.

Culture has been studied in a lab setting by experimentally observing an animal's ability to learn and teach in social contexts, but culture can also be observed in wild populations. These cases are of particular interest to evolutionary biologists: since culture involves the transfer of information between generations independent of genes, it provides an alternative insight into how evolution occurs. Evidence of cultural behavior includes the rapid spread of a novel behavior across a population, complex behavior being taught from a mother to her offspring, and group-specific behaviors that are difficult to explain by other means. There are examples of each of these among killer whales, the species that has become a prime example for culture in nonhuman animals. The communities, clans, pods, and matrilines among resident killer whales are different social groupings that share the same environment and interbreed, ruling out the other explanations for behavioral variation. For example, acoustic dialects among orcas are learned within the matriline and shared by pod members (a group of closely related matrilines). However, among pods, dialects differ, and the differences become even greater between clans (groups of pods). Among communities—for example, between Northern and Southern Residents—there is no overlap in acoustic dialects at all.

Horizontal cultural transmission is social learning that occurs between members of the same generation. This is often seen with new behaviors that can spread rapidly among a population. A classic example of this type of cultural learning in other species is the potato-washing behavior of macaque monkeys in Japan. One macaque figured out that if she dipped her potatoes in water they got cleaner than if she just brushed them off by hand. Other macaques learned this behavior from her and began teaching it to others. Now all the monkeys in this population wash their potatoes. Observations of similar novel behaviors spreading through a population have been made in many other species, including orcas. In Alaska the orcas of Prince William Sound have learned how to take fish off longlines; this ability was learned by all the members of the pod and then transferred to another pod the innovators came into contact with.

Vertical cultural transmission occurs between generations, such as mother-offspring teaching. Since killer whales specialize in particular prey types, they have adapted specific foraging behaviors that can be unique to each population. Calves learn these behaviors from their mothers, and the most well-documented case is in the Southern Hemisphere, where orcas intentionally strand to take seals and sea lions off sandy beaches. Mothers engage in beaching play with their calves and assist them in making their first kills using this tactic, clearly instructing their calves on how to safely engage in this specialized and potentially dangerous foraging technique.

These and other studies looking at the existence of culture in nonhuman animals supplement what I have observed over fifteen years in killer whales: they teach and learn, play and communicate, and have personalities and emotions. We have to be careful not to anthropomorphize too much, by assigning orcas human characteristics where they don't belong. Given what we know about killer whales and their cultural lives, however, there are cases where I feel it is fair to do so. I have taken the liberty throughout this book to describe some of what I have seen in this regard. As humans, we are beginning to realize that despite our very different environments, we have a lot in common with the killer whale.

GREETINGS AND FAREWELLS

Orcas around the world exhibit different behaviors specific to their population: they have unique cultures. One such behavior for the Southern Resident killer whales is the greeting ceremony. It occurs when two groups of whales that have been separated from one another for a period of time reunite. Often, this occurs in the spring when the pods may have been apart for a more extended period, but other times it happens after a separation of only a day. Since the greeting ceremony doesn't take place at all pod "meetings," it isn't clear what situations dictate that one should occur.

When a greeting ceremony transpires, the two groups of whales line up at the surface facing one another. This is a spectacular sight for whale-watchers, as there can be dozens of dorsal fins close together at the surface at the same time. As the whales approach each other, they all dive in unison, and for several moments all appears quiet from the surface. Then, the orcas erupt into surface behaviors such as breaches, spy-hops, and tail slaps, and there is a lot of tactile interaction as they roll around close together. Underwater, via a hydrophone for a human listener, some of the most boisterous vocalizations are heard: it sounds like everyone is talking at once in an unending stream of squeaks, screeches, and whistles. This period of excited social behavior continues as the two groups begin to travel together in the same direction. It's easy to speculate that the whales are celebrating being back together after a time apart, which may very well be what's happening, but greeting ceremonies appear to be a complex behavior. Social play is definitely part of it, but so is sexual behavior, so perhaps these ceremonies play a role in mate selection or other precursors to breeding. Several observations of greeting ceremonies have also raised the question as to whether their occurrence really is restricted to "greeting" or if they may also function as "goodbye" or parting ceremonies.

Researcher Rich Osborne wrote to me about what he called a "separation ceremony" that he observed in 1978. Sitting alone on the shoreline overlooking Haro Strait, he watched as a superpod slowly moved north in two distinct, mixed-pod groups. The whales were close to shore, active, and, to his amazement, totally silent underwater. After half an hour, the two groups stopped and began to circle clockwise, gradually dispersing into their subpod groups. Eventually, Js, Ks, and the L12 subgroup started to head south, while the rest of L-Pod started to head north. In between the two groups were L15 Gracie and L20 Trident, a mother and son. A few short vocalizations chirped over the hydrophone, and both the southbound and northbound groups turned and converged on these two motionless whales. For about twenty minutes the entire Southern Resident community circled clockwise around these two whales, leaving about a fifty-yard buffer around them. Slowly, the group dispersed and spread out across Haro Strait, going their separate ways. "The beauty of this ceremony," Osborne wrote, "was the way these two whales appeared to be the center of it, and the way the other groups of whales that were alternately approaching and retreating from them did so in such a coordinated fashion. The significance of L15 and L20 in all this remains a mystery."

The one clear greeting ceremony I have observed briefly had a similar moment, where the L2 matriline was surrounded by K-Pod. It didn't last twenty minutes, but it included a small group of whales at the center of things, surrounded by whales from another pod. This may be a total coincidence, or perhaps it's our first clue to deeper meaning behind these ritualized ceremonies. At least twice, greeting ceremonies have occurred in the presence of sick whales preceding their deaths. In 1981 a greeting ceremony was observed shortly before the disappearance of five-year-old male J15. In 2005 a greeting ceremony occurred off the west side of San Juan Island on the last occasion that L32 Olympia was seen. Perhaps this behavior occurs whenever the community of whales comes together, whether it be when the pods meet up after an absence or when a member of their tribe passes away. When a whale dies, we don't often know why or what happened. Death often seems to occur when the whales are away from inland waters, where the bodies quickly sink and disappear. Occasionally, a whale is observed in poor condition, or a body will wash up, or we see a mother carrying the body of a stillborn or miscarried calf for several hours on her rostrum. In most cases, however, a whale simply doesn't show up with its family. After several encounters with the family group and without the whale, observers presume that whale is dead. Naturalists cross it off of the orca lists, and that's that: the once vibrant existence of the whale is reduced to a small gray box on our genealogy charts.

When an orca mother loses a young calf, she will sometimes carry it on her rostrum for hours. L72 Racer was observed in September 2010 with her presumably stillborn female calf for more than six hours.
Photo copyright Robin W. Baird/Cascadia Research.

As scientists, as observers of nature, we're taught to maintain an objective distance from our subjects, but obviously we form emotional connections to these animals. It's a rare situation here with the Southern Residents, where the same animals in the same small population return to the same area year after year. It's not just a string of anonymous whales swimming by the coastline during migration, which is how many baleen whales are known. These orcas are individuals whom we have come to know—they have personalities, they have families. Each summer when they return, it's like greeting old friends. Indeed, I have known most of these whales longer than I have known some of the people closest to me in my life. Most of us feel a sadness, even if it is a quiet one, whenever one of the whales dies. We are afraid for them, an endangered population, in an ocean so full of depressing trends and challenges heading into the future (part 3 of this book explores these issues further). I'm sure they know better than us the threats to their survival.

Take, for example, the death of J11 Blossom. She was one of the first J-Pod whales I knew to pass away. I was on a boat near Hein Bank, during an end-of-season crew trip when I heard the news that Blossom hadn't been seen in a few weeks. "What? Are you sure?" I asked. "I thought I had a picture of her just last week." I felt like I had been punched in the gut. Blossom. Could she really be gone? That night at home I looked at my photo again. I had been mistaken—it wasn't her. I felt the loss deeply, as if she were a friend I'd see on the island every summer.

As shocked as I was at the news, imagine what the loss might have been like for Blossom's three offspring: Blackberry, Tsuchi, and Mako. Blossom had been the central point of their lives from the moment they slid from her womb into the ocean. They spent their first year rarely more than a body length away from her. Her strength literally helped pull them along as they traveled up to a hundred miles a day, swimming in her slipstream. Through the first years of their lives, through adolescence, and into adulthood, they stayed physically closer to her than any human children do to their mothers. More often than not, they swam as a family unit, Blossom and her offspring in formation, almost touching. Not only did she teach them how to eat and communicate, like human mothers do for their children, but in this matrilineal culture orca mothers are likely their children's link to the rest of the whale society. She taught them their roles and introduced them to playmates and future mates.

In addition, Blackberry probably understood, even if we can only speculate, that the odds were against him. Adult males are much more likely to die after their mother passes away. Tsuchi may have felt worried and scared. Who would help her through the birth of her own first calf? Who would teach her how to be a mother? Who would babysit? And little Mako, perhaps it was hardest of all for him. The youngest, he was still the closest to his mother, more frequently by her side than anywhere else. Just five years old at Blossom's death, Mako must have felt awfully alone, with a lot to learn still in front of him. Whatever happened to her surely reverberated through the rest of the resident community. She was only thirty-six, in her prime, too young to have died of old age. Among these three pods, every whale knows every other individual, and I have no doubt that they feel loss and grieve, in their own ways.

The fact that orcas grieve got worldwide attention in the summer of 2018 when J35 Tahlequah lost her calf shortly after birth in late July. In an unprecedented vigil, she carried her deceased daughter for at least 17 days. Pushing the body with her rostrum or carrying it in her mouth, Tahlequah refused to let go of her baby until the body began to decompose. The photos of her with her calf made international headlines and brought worldwide attention to the plight of the Southern Residents, eliciting a wide range of emotional responses from humans who learned her story. From one perspective, Tahlequah in her grief did more to bring attention to the issues facing the Southern Residents than we humans did collectively in the decade following their endangered listing.

THE IMPACTS OF CULTURE ON CONSERVATION

Killer whales are capable of social learning and have culture, which can have both positive and negative consequences in terms of their survival. The fact that they can invent and teach novel behaviors in rapid fashion (horizontal cultural transmission) means that they're capable of adapting to the latest situations, such as potentially refining new foraging techniques. But the fact these whales are culturally conservative and tend to do things the way their mothers do (vertical cultural transmission) can be maladaptive. For instance, this keeps them feeding on declining food sources and returning to deteriorating habitats. The size of the Southern Resident population over the past forty years has been correlated with abundance of coastwide Chinook salmon, their preferred prey. When the Chinook numbers are low, the resident whales suffer more deaths, instead of adapting to eat another, more readily available prey species. We know that they are physically capable of catching and eating sockeye salmon, or as I learned by watching Blackberry's family even harbor porpoises, but their culture restricts them from doing so. Why are killer whales so culturally conservative? It may come back to their incredibly strong social ties. Doing what your mother does—synchronizing your behavior to hers—allows you to learn from her and further cements your strong social bonds to her. We see synchronization not only in killer whales but across dolphins, and not just in behaviors like foraging and vocalizing but also in breathing. It's not uncommon to see orca families synch their surfacings to breathe, even when they're not right next to each other.

Killer whale culture has undoubtedly contributed to the evolutionary radiation of orca ecotypes around the world; as populations specialize, particularly via cultural dietary preferences and associated foraging behavior, they diverge from one another into species or ecotypes. Killer whales' propensity to stick to their own culture has led to the diversity we see among the species-complex today known as *Orcinus orca*. Populations of top predators tend to be small, and killer whales make their populations even smaller by further specializing into distinct cultures. These small, specialized populations are more prone to extinction, as they don't have the population size or adaptability to survive changing environmental conditions. Indeed, some scientists have theorized that the Southern Residents may be a cultural remnant of a larger killer whale population that is becoming less viable at the more southern latitudes as global climate changes and Pacific salmon populations decline. Although we very much want to keep them around for the immediate future, in the longer term the Southern Residents may be just one of many killer whale ecotypes that come into existence for a while to eventually be replaced by another orca culture better adapted to live in a slightly different world.

We recognize the importance of preserving human cultural diversity, but we are just beginning to recognize the importance of preserving cultural diversity as a type of biodiversity. The endangered listing of the Southern Residents in the United States was first denied because killer whales as a species were seen to be stable in the northeast Pacific. It wasn't until they were identified as a distinct population segment in the mid-2000s, in large part due to their cultural uniqueness making them a separate breeding population, that they received protection under the Endangered Species Act. The Southern Residents are recognized as a distinct culture of killer whale, and their protection lays the foundation for other subpopulations to be protected based on their unique cultures as well.

The way humans treat killer whales and other cetaceans may transcend traditional environmental protections as we begin to recognize the rights that might be owed to a fellow cultural species. There's been an effort by some animal rights activists to recognize orcas and other highly social, highly intelligent species as nonhuman persons, with the understanding that our morality urges us to treat all persons with certain respect. In a philosophical sense, a "person" is not a term synonymous with "human" but rather is used for any entity worthy of moral standing because that entity has a conscious mind in addition to a body. At the 2012 annual meeting of the American Association for the Advancement of Science in Vancouver, BC, there was a panel discussion on the ethical issues that arise from the latest science showing the intelligence and culture of whales and dolphins. Also presented was a Declaration for the Rights of Cetaceans, which calls for basic rights to be bestowed upon all cetaceans, including the rights to life and freedom as well as the protection of their environments and cultures. Supporters of the declaration include the international group Whale and Dolphin Conservation and the United States–based Nonhuman Rights Project. In 2013, India became the first country to recognize dolphins as nonhuman persons, banning them from being held captive at any facility in the country. Widespread acceptance of such ideas is likely still a long way off, but it is undeniable that going forward, culture will play a larger role in conservation issues.

LEARN MORE

For more on culture in cetaceans, see Carl Safina's *Beyond Words: What Animals Think and Feel*, Hal Whitehead and Luke Rendell's landmark paper "Culture in Whales and Dolphins" and their book *The Cultural Lives of Whales and Dolphins*, as well as journal articles by scientists such as Lance Barrett-Lennard and Rüdiger Reisch.

PART 2

OUR CHANGING RELATIONSHIP WITH THE SOUTHERN RESIDENTS

CHAPTER 5

The Salish Sea
Where Orcas and Humans Come Together

EIGHTEEN THOUSAND YEARS AGO, DURING the last ice age, the Pacific Northwest region looked entirely different than it does today. The Arctic ice sheets were advancing south, filling in the plains that stretched between the Cascade and Coastal mountain ranges. So much of the world's water was locked up in glaciers that the sea level was three hundred feet lower. The round hills and rugged valleys had yet to be carved out by ice and eroded by meltwater. There was no such thing as Puget Sound or Vancouver Island, just one connected land mass.

By fourteen thousand years ago, the ice had reached its southernmost advancement, just south of modern-day Olympia, Washington. Near what is now the United States–Canada border, the ice sheet was over a mile thick. It completely covered the land masses that would become the San Juan and Canadian Gulf Islands, with one lobe stretching south into what would become Puget Sound and the other west to the Pacific Ocean, running through what would become the Strait of Juan de Fuca. Rivers running beneath the massive ice floe started carving out the deep channels that would become the straits of a new inland sea. The climate shifted, warmed, and the ice began retreating at a faster pace than it had advanced. As the ice receded, it further shaped the landscape, scraping off the tops of the mountain peaks it passed into rounded summits and leaving behind deposits of glacial till and large boulders known as glacial erratics that had been carried south from the Canadian mountains. The land rebounded from the tremendous weight of the ice and simultaneously the sea level rose dramatically as glaciers turned into meltwater.

By ten thousand years ago, the local geography looked much as it does today, with Puget Sound and the Georgia Depression filling with water, leaving just the tops of former hills and mountains as islands above their surface. This new habitat was ideal for Pacific salmon, which through the ice age had found refuge further south in the Columbia–Snake River basin. As the ice receded and the inland sea filled, the salmon could expand their range to the north, where they populated Puget Sound and the Fraser River. It wasn't an easy task adapting to this region, which was constantly shifting as the remnant ice melted and the global climate warmed. The flow of the rivers and streams was always changing, but the salmon survived because of their genetic diversity and their tenacity. They gained a solid foothold in the emerging Pacific Northwest ecosystem. By four thousand years ago, both the climate and the river routes had stabilized, and the lush fir and cedar forests that we are familiar with today grew along the banks of the salmon rivers.

THE PEOPLE OF THE LAND AND THE PEOPLE OF THE SEA

Two nations followed the salmon to this inland sea: terrestrial (the Coast Salish peoples) and marine (killer whales). The Coast Salish peoples were a group of more than fifty interrelated tribes that moved west, separating from their interior cousins. The different groups shared a common linguistic and cultural ancestry and lived a life well suited to the world around them. They were hunter-gatherers with permanent winter villages and temporary summer settlements, relying on the evergreen forests around them, particularly western redcedar, for building their longhouses and dugout canoes. Salmon was central to their diet, and they honored the first spawners that returned every spring in annual first salmon ceremonies. The homeland territories of the Coast Salish peoples matched the rivers and creeks the salmon runs used for spawning, with different tribes utilizing different watersheds. They also relied heavily on shellfish and berries, and cultivated the grassland prairies around them to yield other roots and tubers such as camas bulbs. A patrilineal people, relationships between the Coast Salish were made and strengthened via marriage and trade, with potlatches being a central component of community life. They believed in the oneness of all living things and considered the animals around them as different types of people with whom they shared the world, as other nations worthy of the utmost respect.

Beginning ten thousand years ago, the marine nation of killer whales followed salmon to the inland sea. Like the terrestrial nation of Coast Salish peoples, the marine nation

A Lummi canoe pauses to watch an orca. For thousands of years, Coast Salish peoples and the resident killer whales have lived side by side in the Salish Sea.

also had multiple tribes (pods in their case) that shared a linguistic and cultural heritage with salmon at its center. They came and went from the region, following the fish, but feasted in the inland sea during the summer months when the salmon were most plentiful. A matrilineal society, the orcas' world revolved around the mothers and grandmothers of the pods, who carried the most knowledge. Familial relationships were strongest, but community events such as greeting ceremonies and gatherings of extended families into superpods strengthened the links between the different groups. As we know from tribal legends, the relationship between the Coast Salish peoples and the killer whales was one of mutual respect. Both were nations of expert hunters, and there was an unspoken agreement within each nation that they would not hunt each other. Among the

Coast Salish, to kill someone in another tribe would be an unspeakable act of disrespect, and it was rarely done. There was enough fish for everyone, and no need for any single tribe to take more than it needed. The groups weren't competitors but compatriots who thrived off the bounty of this coastal ecosystem.

Coast Salish scholars tell us of an old legend among the Coast Salish peoples called the Killer Whale and the Thunderbird. One time, long ago, the salmon rivers suddenly went empty, and there were no fish for the people to eat. They traveled down the rivers to identify the source of the problem, where they found Killer Whale eating all the salmon before any could travel up the rivers. Coast Salish people paddled their canoes out to try and chase the whale away, but he was too large to be deterred. All the medicine people from the different tribes joined together to sing ancient spiritual songs for four days and four nights, calling for help from Thunderbird. At the end of the fourth day, according to the legend, Thunderbird appeared and battled Killer Whale. Thunderbird was victorious and flew off with Killer Whale in its talons, after which the salmon returned to the rivers for the people. Ever since, killer whales would eat only some of the salmon, leaving enough to travel up the rivers to feed the Coast Salish peoples and other animals.

Just over two hundred years ago, however, the balance that had been maintained between the terrestrial nation and the whale nation was disrupted with the arrival of the first European explorers. In late 1700s the Europeans arrived along the Northwest Coast, including Captain George Vancouver and the British. These explorers brought with them many new ideas and inventions that wrought inevitable changes. Novel types of weapons as well as never-before-seen colorful fabrics and beads were traded in exchange for luxurious furs the Europeans shipped back home. One of the consequences of their arrival was the transmission of new communicable diseases, including smallpox, that ravaged the Coast Salish people, decimating the population by almost two-thirds in the initial one hundred years after the European's arrival. As trading outposts, forts, and cities were established, the white settlers altered the landscape in devastating ways, including clear-cutting to feed the colonizers' ongoing growth and replacing thriving forests with cultured fields and pastures to yield more familiar crops. The Coast Salish peoples found ways to adapt, however. They were hired by the settlers to help log and farm and, given their expertise, to fish. Community-sustaining ceremonies such as potlatches were eventually banned by the colonizers, and piece by piece many Indigenous communities began to lose their culture.

For the most part, the marine nation of killer whales along the Northwest Coast was overlooked upon the arrival of the Europeans. The community was too small to be of interest to commercial whalers, and except for being occasionally shot at by white fishermen, the orcas were left to their own devices. The arrival of the colonizers ushered in the concept of humanity's exploitation of and domination over nature, which over time certainly affected the killer whales. As canneries sprung up along the West Coast, the Pacific salmon the whales depended on began to dwindle. In a reversal of that old Coast Salish legend, by the mid-nineteenth century it was now people eating all the fish, leaving none for Killer Whale.

THE NAMING OF THE SALISH SEA

Puget Sound, the Strait of Georgia, and the Strait of Juan de Fuca are all distinct on a map, but they are in fact connected waterways. From the Canadian Gulf Islands through the San Juan Islands down past Seattle into the depths of Puget Sound, these areas are all influenced by the waters that surge in and out with the changing tides around Vancouver Island's south end. For many years this geographic region was referred to by the cumbersome name of "the Puget Sound–Georgia Basin." From a biological and ecological perspective, however, it makes more sense to recognize this inland sea as a single entity, although the effort to do so spanned several decades.

In the 1970s marine biologist Burt Webber of Western Washington University was studying the risks of an oil spill in the region because plans were in place to send freighters of crude oil from Alaska to refineries in Washington. Through these studies, Webber realized that the inland waters of Washington State and British Columbia were part of a unified, but unnamed, ecosystem. He recalled one meeting during which a government official referred to the waters off Bellingham, Washington, as "Northern Puget Sound." This just didn't seem right to Webber. The idea of the waters being part of a single ecosystem became clearer when he realized the entire inland sea is a single estuary. The salty waters of the Pacific Ocean mix with the freshwaters of the region's rivers, resulting in a productive

The Salish Sea, which encompasses the Strait of Juan de Fuca, Strait of Georgia, and Puget Sound. Map by Sara Hysong-Shimazu.

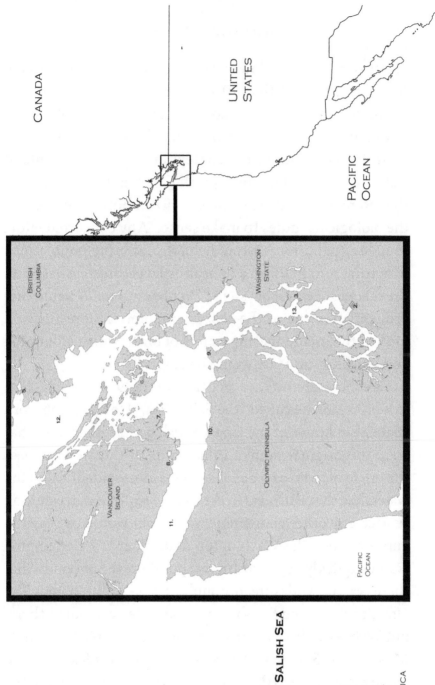

LOCATIONS IN THE SALISH SEA

1. OLYMPIA, WA
2. TACOMA, WA
3. SEATTLE, WA
4. BELLINGHAM, WA
5. VANCOUVER, BC
6. FRIDAY HARBOR, WA
7. VICTORIA, BC
8. SOOKE, BC
9. PORT TOWNSEND, WA
10. PORT ANGELES, WA
11. STRAIT OF JUAN DE FUCA
12. STRAIT OF GEORGIA
13. PUGET SOUND

ecosystem that has lower salinity than the open ocean. Research revealed that 70 to 80 percent of the freshwater to the inland waters comes from British Columbia's Fraser River. Webber realized that looking at the salinity of Puget Sound in the context of Washington's inland waters alone wouldn't be logical. One had to look at all of the interconnected waterways between both countries for the biological studies to make sense. "Any concerted effort to reduce pollution, restore habitat, and bring populations of marine resources back to healthy levels must be based on the science of the ecosystem," Webber wrote. "A name that brought attention to the ecosystem interrelationships and unified the inland marine waters of both British Columbia and Washington State would be helpful."

Webber submitted a petition in 1989 proposing that this bioregion be called "the Salish Sea." He chose the word "Salish" in honor of the Coast Salish people, the tribes that were the original human inhabitants of the area. Indeed, the range of the Coast Salish tribes is almost identical to the watershed that all drains to the Salish Sea. The term wouldn't replace any other geographic names but would overlay the entire region. The Washington State Board of Geographic Names quickly rejected the proposal on the grounds that there was no widespread popular usage of the name. The effort to name the Salish Sea could have easily died there, but Webber's idea did attract some supporters. The Whale Museum of San Juan Island endorsed the idea, and the name made sense to the residents of the Gulf Islands and the San Juan Islands who felt like there was no cohesive name for the waterways where they lived. Educators and land use planners started to use the term, and natural resource managers slowly followed suit. In 2005 the transboundary ecosystem conference that formerly used the name Puget Sound–Georgia Basin began to refer to the Salish Sea.

The first official recognition of the term Salish Sea came in 2008, when the Coast Salish tribes on both sides of the border formally adopted the name to refer to their shared ecosystem. This, along with the increasing use throughout the region, led Webber to ask both British Columbia and Washington State to reconsider his 1989 petition. In 2009, twenty years after he officially proposed the term, both British Columbia and Washington endorsed the name. Shortly thereafter, it was approved by Canada's Geographical Names Board and the United States Board on Geographic Names. The Salish Sea officially ranges east from the entrance of the Strait of Juan de Fuca and extends from the northern reaches of the Strait of Georgia to the southernmost portions of Puget Sound, about seven thousand square miles. It includes all the land that drains into these waterways, another forty-two thousand square miles or so. Today, more than eight million people live within the drainage basin of the Salish Sea, with most of them in the cities of Nanaimo, Victoria, Vancouver, Bellingham, Seattle, Everett, Tacoma, Olympia, Port Townsend, and Port Angeles.

In addition to these human residents, the Salish Sea directly supports more than 30 mammal species and 170 bird species as well as 260 fishes and more than 3,000 species of marine invertebrates that either live in its waters or rely on its marine resources as a source of food. Many of these species, including killer whales, daily cross the invisible international border.

For thousands of years the Salish Sea has been a place where orcas and humans spend time in close proximity. Those who are lucky enough to get to know the whales individually inevitably develop favorites. J41 Eclipse is among my personal favorites of the Southern Residents. One particularly memorable encounter during the July 4th holiday on San Juan Island was the beginning of my relationship with her. It was early morning at Lime Kiln Point State Park and the tranquil waters mirrored the sky, calm enough that you could hear the harbor porpoises breathing as they surfaced a quarter-mile offshore. I sat at a picnic table with my journal, and a friend of mine was perched on a rock, whittling a piece of wood into the shape of a killer whale. There were a few other people scattered along the shoreline, but all were engaged in similarly personal activities, everyone taking in the beautiful morning in their own way.

In the height of summer, early morning is often the only time of day no one is sure of the whereabouts of the orcas when they are in inland waters. In an era of cell phones, handheld radios, e-mail, text messages, and social media, anyone in the whale-watch community can keep tabs on the movements of the whales from the moment they're found in the morning until the last whale-watching boat leaves them at dusk. Whether this abundance of knowledge is a good or a bad thing is up for debate, but it does take a little bit of the mystery out of the experience. There's something to be said for going out and looking for whales, hopeful but without the certainty of whether you will see them. These days, you can only capture that feeling at dawn, before the boats go out and after the whales have had all night to travel somewhere else. That's what we were doing at Lime Kiln, a couple of young adults drawn out of bed much earlier than we would normally see fit, all for the chance of seeing the orcas.

It turned out Lime Kiln was definitely the place to be. A blow much louder and longer than that of a harbor porpoise broke the silence—a signal to abandon everything, grab the camera, and head to the shoreline. I climbed down to my favorite rock in front of the lighthouse, as low to the water and as close to the whales as I could get. Even as more people gathered on the rocks above me in anticipation, from down at the water's edge I blocked out all other distractions, eager for it to be just me and the whales. In the lead were J1 Ruffles and J2 Granny, a little further offshore and well ahead of the rest of the pod. This meant J-Pod, who had spent the last several days in Rosario Strait, had looped around into Haro Strait during the night. Behind Ruffles and Granny came J8 Spieden, her wheezing blow especially noticeable on this quiet morning. I looked for her granddaughter Shachi, who a couple of days before had been reported with a brand-new calf. J-Pod had been in Rosario Strait the night of June 30, with no calf. First thing the next morning, the newborn calf, designated J41, was tucked in tight to Shachi—it had definitely been born sometime during the night. Everyone who saw J41 on that first day commented on how small she looked, so I was anxious to see if she was still with Shachi. Shachi had lost her first calf, J29, over a decade earlier and was past due to have another.

Sure enough, Shachi wasn't far behind Spieden, and popping up just beside her was the smallest orca I have ever seen. Still the bright orangey color typical of a newborn, J41 had visible fetal folds on her forehead and down her back.

Though she must have been the typical five or six feet long of an infant orca, she was dwarfed by her adult mother, her tiny curved dorsal fin just a fraction of the height of Shachi's. The mother and calf surfaced together three times in front of me, and on the last dive Shachi gave a tail slap, her flukes disturbing the flat-calm water. So small, so vulnerable, yet so full of life, J41 was a symbol of hope for a population on its way to being listed as endangered later that same year.

Most whale-watchers form a special bond with one particular whale, an orca with which they have numerous significant encounters. For me, that whale is J41. I watched the rest of the season as she stayed close to mom, occasionally getting a playful visit from J40 Suttles, who was born just six months before her. I eagerly awaited her return the following summer, as the first year of life is full of perils for a young orca, including the strain of keeping up with their pod from day one. Unfortunately they also inherit an influx of toxins through their mother's milk. J41 came back the next summer having lost her baby pink and grown a lot; she was now the traditional black-and-white of a killer whale. On her first birthday, I saw her alongside Shachi leading a superpod through Haro Strait. Later in the summer, through the Whale Museum's Orca Adoption Program, she was named Eclipse.

A few seasons later, I was on a whale-watching trip when a young calf surfaced by itself behind our parked boat, excitedly lunging in different directions. It took me a moment to recognize Eclipse, now three, who proudly spyhopped with a small fish in her mouth. Shachi soon came over to collect her calf, and the two swam off together. Eclipse was learning to hunt and catch her own fish. Another time I

A days-old J41 Eclipse next to her mother, J19 Shachi, off Lime Kiln Point State Park.

New mother J41 Eclipse with her own calf, J51 Nova, off the same rocks at Lime Kiln Point State Park.

was on a kayak trip with some friends off the west side of San Juan Island. We had no sooner launched our kayaks and pulled into a kelp bed than we were circled by Shachi and Eclipse. Even though she was only a few years old, it's amazing how much bigger Eclipse looked from the perspective of a kayak! In that moment I was definitely a visitor in her world, rather than watching from the shoreline or from the deck of a boat.

Perhaps most memorably, in 2015, I was on the water the day Eclipse was seen with her own first calf. The fairly gray, drizzly February day was brightened by the little black-and-orange orca bobbing to the surface between Eclipse and Shachi. At first researchers assumed it was Shachi's calf, because Eclipse was just nine years old—two years younger than the youngest documented orca mom. But in the coming weeks and months the close association of the calf to Eclipse showed that she had indeed become the new youngest mother on record! The new little calf, a male, was designated J51 and later named Nova. Now I get to watch him grow up as he returns each year beside Eclipse with other Southern Residents to the Salish Sea.

TRAVEL PATTERNS OF ORCAS IN THE SALISH SEA

Killer whales are just one of the species that make use of the Salish Sea in its entirety, using the whole region with no regard for the invisible borders of humans. Like fishermen making return trips to favorite fishing grounds over the years, the whales also have their preferred routes through their core habitat. Throughout the more than forty years scientists have been monitoring their movements, some patterns have emerged in where J-, K-, and L-Pods are likely to be in any given month of the year. Not surprisingly, their location has a lot to do with salmon.

When the days are getting noticeably longer in the spring, it is time for the Southern Resident community of killer whales to begin returning to their summer home in the Salish Sea. J-Pod is almost always the first pod to return, beginning to ply the waters near the San Juan Islands with regularity in March, April, or May, depending on the spring salmon runs. (In recent years, with the further crash of Fraser River spring Chinook runs, this has become more like late May or June.) They usually return to inland waters from the Pacific Ocean through the Strait of Juan de Fuca, traveling east along the southern tip of Vancouver Island and the northern Olympic Peninsula past places like Neah Bay and Port Townsend on the Washington side and Jordan River, Sooke, and Victoria in British Columbia. It doesn't take long for J-Pod to take up their typical cyclical summer travel pattern, which takes them from the Strait of Juan de Fuca to the Fraser River (where many of the main salmon runs are heading) and then back south to the San Juan Islands.

After entering the inland waters, the orcas head up Haro Strait past the west side of San Juan Island toward Turn Point on Stuart Island. Once there, they take one of two routes up to the Fraser River: either north through Swanson Channel and through Active Pass in the Canadian Gulf Islands, or east up Boundary Pass, so named for being the international boundary between the United States and Canada. The whales often spend time foraging near the mouth of the Fraser, just north of Vancouver, before looping back south again. On their way south, the decision point is at East Point on Saturna Island, where the whales will either proceed down Boundary Pass back toward Turn Point and Haro Strait or angle toward Rosario Strait, which will take them down the east side of the San Juan Islands and past Anacortes. Either way, J-Pod eventually ends up back off the south end of San Juan Island, where they have another popular foraging ground between Salmon and Hein Banks.

K- and L-Pods also vary their return to the Salish Sea, presumably due to the seasonal shifts in salmon abundance. They typically show up in May, June, or July, sometimes together or sometimes one after the other. Although they also will often enter through the Strait of Juan de Fuca, on several occasions these Southern Residents have made their way into inland waters by swimming around the north end of Vancouver Island and down through the Inside Passage. This travel route takes them through Johnstone Strait, typical late-summer stomping grounds for the Northern Residents, and down through the Strait of Georgia past Nanaimo to the mouth of the Fraser River. The first time Southern Residents were photographed in Johnstone Strait was May 1994, when the L12s, L25s, and K-Pod were seen there. This travel route became more common after this sighting. For several consecutive years, when arriving in the Strait of Georgia closer to their normal summer home range, at least part of L-Pod

The traditional core summer range of the Southern Resident killer whales. Map by Sara Hysong-Shimazu.

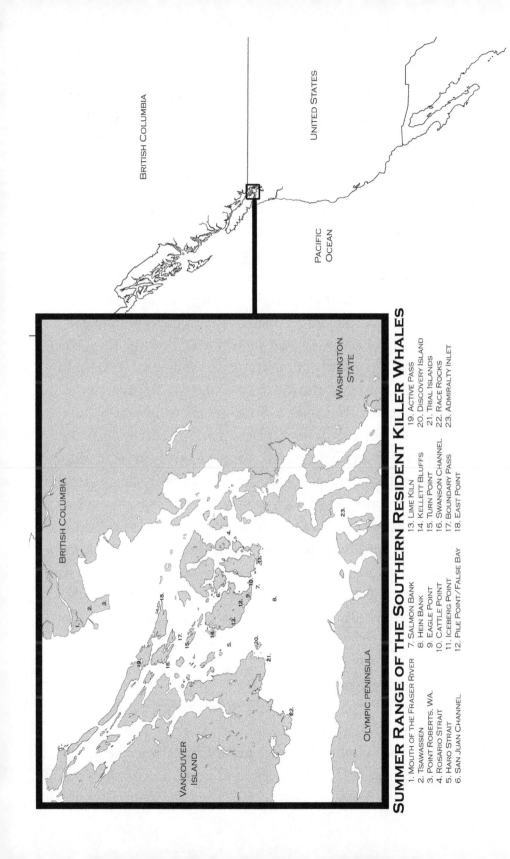

SUMMER RANGE OF THE SOUTHERN RESIDENT KILLER WHALES

1. MOUTH OF THE FRASER RIVER
2. TSAWASSEN
3. POINT ROBERTS, WA.
4. ROSARIO STRAIT
5. HARO STRAIT
6. SAN JUAN CHANNEL
7. SALMON BANK
8. HEIN BANK
9. EAGLE POINT
10. CATTLE POINT
11. ICEBERG POINT
12. PILE POINT/FALSE BAY
13. LIME KILN
14. KELLETT BLUFFS
15. TURN POINT
16. SWANSON CHANNEL
17. BOUNDARY PASS
18. EAST POINT
19. ACTIVE PASS
20. DISCOVERY ISLAND
21. TRIAL ISLANDS
22. RACE ROCKS
23. ADMIRALTY INLET

made an unprecedented trip west through Active Pass and down Swanson Channel, a direction of travel Southern Residents up until this point had only been observed to do in the opposite direction.

July, August, and September are the heyday for Southern Residents in the Salish Sea. The pods travel their regular circuit in various social group combinations, often remaining together as a superpod for days at a time, meaning as many as eighty or more whales are traveling together. The whales will travel, forage, socialize, and rest all throughout the Salish Sea, although they often engage in certain behaviors in particular areas. For instance, the whales will regularly spread out and forage near Salmon and Hein Banks off the southwest side of San Juan Island; exiting Active Pass, meanwhile, seems to be party time, as the whales regularly get very surface active with the whole group breaching, tail slapping, and cartwheeling. Periodically, some or all of the whales will leave inland waters for a period of days or weeks, heading back out to the mouth of the Strait of Juan de Fuca and the Pacific Ocean to feed on salmon there. L-Pod spends the most time out at the ocean, although it isn't unheard of for all three pods to head west together. One memorable summer in the 1990s, all three pods were out of inland waters for a total of five weeks before returning, a rare occurrence in those times; sadly, these long absences have become more common in recent years.

When all the whales are in the Salish Sea, a pod or sub-pod seems to be designated to patrol the west side of San Juan Island in what local whale-watchers describe as the "west side shuffle." The whales head north in Haro Strait to a point such as Land Bank's Westside Preserve, Lime Kiln Point State Park, or Kellett Bluffs on Henry Island, and then they flip to travel back south. They swim south past False Bay, Eagle Point, and down to Salmon Bank before turning again and heading back north. Even when some of the other whales are traveling north to the Fraser River, this group of whales will go back and forth as many as three or four times throughout the course of a summer day. Throughout the last few decades, the "west side shuffle" group has variably been K-Pod, the L12 subpod, and J-Pod.

J16 Slick and her daughter, J42 Echo, wow onlookers at Lime Kiln Point State Park.

By the end of September, resident orca sightings around the San Juan and Gulf Islands tend to dwindle, although some years the whales remain consistent in their travel patterns well into October. As the days shorten and fall arrives, the whales typically head south into Puget Sound, a place they rarely travel in the spring and summer months. After passing through Admiralty Inlet, the Southern Residents will go past the southern end of Whidbey Island, past Seattle, as far south as Vashon Island. During this time of year, diet studies from prey and fecal sampling have shown their diet shifts from primarily Chinook to including a large percentage of chum salmon. By the end of the calendar year, the salmon numbers have decreased, so the orcas leave in search of food elsewhere. For many years it was unclear where they traveled, and while winter sightings remain scarce, a picture of the winter feeding grounds of Southern Residents has started to emerge. All three pods head out to the open ocean and roam further north and south than they do during the rest of the year, likely taking advantage of Chinook salmon populations that congregate coastally near the continental shelf rather than migrate further north and offshore as some other salmon species do.

Js, Ks, and Ls travel as far north as Haida Gwaii (formerly the Queen Charlotte Islands off central British Columbia), and it is assumed they spend a lot of time off the Washington and Oregon outer coasts feeding on Columbia–Snake River salmon. On January 29, 2000, K- and L-Pods surprised researchers by showing up off Monterey, California, extending the known range of Southern Residents by more than five hundred miles. Proving this wasn't a fluke occurrence, members of both pods were also seen in the area at least once during January and February most years since then, the first indicators researchers had that the Southern Residents were using the outer coast from Washington all the way to California. This California travel pattern seems to be unique to K- and L-Pods. J-Pod has only been recorded once on the Cape Flattery hydrophone and never on the hydrophones south of there, suggesting that their outer coast range extends more to the north of the Strait of Juan de Fuca.

J-Pod spends the most time in the Salish Sea, almost as if they are somehow the dominant pod (although no signs of aggression between the Js, Ks, and Ls have ever been witnessed). Js can be seen regularly in the Salish Sea year-round. K- and L-Pods seem to join in during the summer and fall months when salmon are more abundant, but in the winter they travel the outer coast instead. They have to expend a lot more energy than the Js, which may mean they metabolize more of the toxins in their blubber, and because of where they travel, they are also exposed to different toxins. As a result, J-Pod has historically been the healthiest of the three pods, with the highest survival rate among all age-sex classes; Ks and Ls have struggled more, particularly with calf mortality.

It's interesting to speculate about these changing travel patterns over the years and how they might relate to some type of social order among the three pods. K-Pod, for instance, used to travel with J-Pod much more regularly than today. Perhaps with the death of their oldest matriarchs, K7 Lummi and K11 Georgia, they lost some of their strong social ties to J-Pod's elders. After the deaths of these two whales, it certainly seemed like K-Pod started spending less time in inland waters. Certain L-Pod matrilines seem to hold different statuses. The L12 subgroup was notorious for spending a lot of time on the west side of San

Juan Island in the summer, indicating that perhaps they are "in" with J-Pod more than the rest of the Ls. In recent years, perhaps after the death of matriarch L12 Alexis, that has shifted somewhat and the L12s have become less frequent summer visitors. Other matrilines have occasionally traveled with J-Pod in the summer, including in various years the L7s, K14s, and K16s. This indicates that for some reason these matrilines may have been granted a different status than the rest of their pod. The L2, L5, and L54 matrilines, by contrast, are far more irregular visitors than the rest of L-Pod, some summers only visiting inland waters two or three times over the entire course of the year.

The summer of 2013, when the Southern Residents spent very little time in the Salish Sea, is another interesting example. J-Pod was late in returning in the spring, and was only scarcely seen in the months from May through August. Instead, the Salish Sea got regular visits from various L-Pod subgroups. The theory was that the salmon were most abundant offshore during this summer, and that this is where J-Pod was hanging out. (When Js did come inland, they looked robust, indicating that they were probably getting enough to eat somewhere.) Perhaps with the food being distributed differently this year, the travel patterns changed accordingly. J-Pod now spent most of the summer away from the Salish Sea where the salmon were, while L-Pod was "allowed" to come into inland waters regularly to see what was available to eat there. The L54 subgroup, usually rare visitors and perhaps the lowest "ranked" Southern Residents when it comes to Salish Sea fish rights, came in often in 2013. For one stretch in August they were the only whales there, regularly patrolling the west side of San Juan

Island in the way the L12 subgroup usually did with J-Pod during better Chinook salmon years. This is all speculation, but it seems plausible that if different orca populations have access to different feeding areas, these unspoken agreements may also extend to other levels of orca social structure such as pods and matrilines.

While a typical year for a Southern Resident whale thus consists of making the circuit from the San Juan Islands to the Fraser River in the spring and summer, spending time in Puget Sound in the fall, and roaming the outer coast in the winter, there are of course aberrations to the typical traveling routes. Several times each year, some of the Southern Residents head either north or south through San Juan Channel, the narrow waterway that passes right by the town of Friday Harbor. On one July afternoon I was working on the *Western Prince* when we encountered a portion of L-Pod right off the south end of San Juan Island. They were close to shore doing a lot of socializing. We were following L105 Fluke and his uncle L95 Nigel when the strong flood tide seemed to push them into the heavy currents of Cattle Pass at the southern end of San Juan Channel. Fluke breached a couple of times, and it wasn't long until the other twenty or so whales followed the first two into the pass. On our morning trip earlier that day, I had committed the digital photographer's cardinal sin of not having a spare camera battery when mine died. As I watched this unusual event unfolding, I had the feeling we were about to witness something special.

The underwater geography of Cattle Pass includes some extreme depth changes, and with the up to twelve-foot tidal exchange in inland waters, the narrow channel can make for some intense and unpredictable currents, complete with upwellings, back eddies, tide rips, and whirlpools. At times, Cattle Pass can look more like a river with running rapids than a saltwater channel, and on some days we would even stop the boat to let the passengers experience the phenomenal power of the currents as they spun our forty-six-foot vessel with ease. On this day, it simply looked like L-Pod was playing in the tumultuous water. They were being pushed by the tide as they hung at the surface, rolling around, slapping their fins, and doing lots of spyhopping. Right before heading home (and when some of our passengers began to complain of dead camera batteries too, from having snapped so many photos over the previous hour), we cut the engines to take one last quiet look. Suddenly a group of three whales we hadn't seen before surfaced off our bow: L26 Baba, L90 Ballena, and L92 Crewser. They swam the entire length of the *Western Prince*. The lighting was just right so that you could see their bodies underwater, and one of whales, Crewser, was swimming upside down, his flukes breaking the surface in a lazy tail slap. Ballena turned her head as she went by, unmistakably looking up at us. I remember it vividly, in part because for once, without a camera in hand, I could do nothing but simply take it in. I often would share pictures I took with passengers after a trip, but this was one encounter where I asked them to send their photos to me instead!

Occasionally the whales show up in waters they rarely visit or have never been documented in before. One such incident began on the morning of October 21, 1997, when nineteen whales from L-Pod (the L4, L21, L26, L37, and L66 matrilines) navigated the Port Washington Narrows in Puget Sound and entered Dyes Inlet, marking the first time orcas had been seen in the inlet in at least thirty to forty years. As the days passed and the whales remained in Dyes Inlet, they attracted more attention as whale-watchers flocked to the shoreline and out onto the water to get a closer look. As the days turned into weeks, people began to question if the orcas might be "stuck"; they repeatedly approached their only exit—the Port Washington Narrows—but kept turning back before passing under the Warren Avenue Bridge. Things came to a head on November 9, when an estimated five hundred boats and kayaks were on the water in the narrow inlet and the whales started to show signs of stress similar to when they were trapped in bays during the capture era. The animals "paced" the inlet and were no longer vocalizing or foraging during the day. Some whale researchers monitoring the situation advocated for some type of rescue effort, but the idea that the whales were stuck was debated, as the water under the bridge was twenty-five feet deep—plenty deep for the orcas to depart on their own terms.

On the drizzly morning of November 19, with only a handful of boats present, the whales finally made their move. They approached the bridge in two groups, led by L21 Ankh, but turned back and proceeded to mill for about ten minutes. Again, Ankh led her group toward the bridge, but once more, she turned back. Ankh was the oldest matriarch present at forty-seven years old, and she seemed determined. After some more back and forth, she approached the bridge with fellow matriarch L26 Baba, the second oldest whale in the group at age forty-one. This time, they did high-arch dives like they were going down deep, and a moment later they surfaced on the far side of the bridge, breaching. The other whales followed their lead, passing under the bridge in twos and threes. Once clear of the bridges, the L-Pod whales porpoised back into the more expansive waters of Puget Sound, quickly swimming away from where they had spent the last month. The only other similar incident where Southern Residents stayed in one place for an extended period was in 1991, when J-Pod spent eleven days seemingly "stuck" in Sechelt Inlet near Vancouver. They also showed pacing behavior there before porpoising back out their only exit near midnight on a slack tide.

One might have expected that after such a potentially stressful incident in Dyes Inlet that the whales would have steered clear of Puget Sound and its tangle of inlets after that. In fact, their use of Puget Sound in the autumn months has actually increased since then. In the five years preceding the Dyes Inlet incident, all three pods were seen in inland waters in only three of the fifteen fall months (October–December, 1992–1996). In the five years after the incident, all three pods were seen in inland waters thirteen of the fifteen fall months (October–December, 1998–2002). Perhaps this incident actually helped the resident whales discover the benefit of feeding on fall salmon runs in Puget Sound, although they have yet to return to Dyes Inlet itself.

Ever since the ice retreated and the salmon arrived in what is now modern-day Washington and British Columbia, the people of the land and the people of the sea have continued to live side by side. The relationship between the terrestrial human nation and the marine whale nation has been an ever-changing one, with the killer whales having to adapt as the ever-increasing human population puts more stress on the environment. The Southern Residents have made some shifts in recent decades over where they travel and how often they visit regional waters, presumably in response to where they are able to find food. Just as in ancient times, the salmon rivers are beginning to run empty, but this time there will be no Thunderbird to come and restore balance. Instead, it is up to humans to decide if we are going to eat all the salmon or leave enough for our neighboring animal nations to continue to live alongside us in the Salish Sea.

LEARN MORE

For more on the history and ecology of the Salish Sea, read Steve Yates's *Orcas, Eagles, and Kings: The Natural History of Puget Sound and Georgia Strait*, and Audrey Benedict and Joseph Gaydos's *The Salish Sea: Jewel of the Pacific Northwest*. For more on the Dyes Inlet incident of 1997, see the contemporaneous newspaper articles that ran in the *Kitsap Sun*.

CHAPTER 6

The Capture Era

IN THE SUMMER OF 1964, student David Jamison was at the University of Washington's Friday Harbor Labs studying limpet behavior and helping a visiting professor survey fish in tide pools. He spent a lot of time along the shorelines of San Juan Island and occasionally was thrilled by sightings of killer whales. One day down at Cattle Point, he saw a small group of whales pass just ten feet away from the cliff below him. Momentarily forgetting the tide pools he was heading toward, Jamison instead ran up the hill to watch the whales round the corner by the lighthouse. As the orcas made their way out into the straits, in their path was a small fishing vessel with two men aboard. Jamison watched as the fishermen moved away from the sides of their boat and squatted together in the middle of the deck, nervously keeping an eye on the four "blackfish" (the regional fishermen's term for killer whales) approaching them. The whales took another breath and then dove, passing harmlessly under the boat and continuing on their way.

On another afternoon, Jamison, his wife, and their young child were enjoying a picnic near the Lime Kiln

Lighthouse. After spotting a dark, triangular dorsal fin splitting through the water, the family moved down onto a rocky point just south of the lighthouse to watch the whales swim by. The orcas approached from the north, one of the larger groups Jamison had seen. Camera in hand, he snapped a photo of a male and female whale surfacing together not far from where he stood with his family. Out of the corner of his eye, Jamison saw someone else walking toward the lighthouse. He didn't pay much attention to the figure at first, assuming it was the lighthouse keeper coming down to see the whales, but a moment later an unexpectedly loud cracking noise interrupted the peaceful moment. Jamison whipped around to see the man in front of the lighthouse with a gun held in front of him, aiming down into the water.

It took a moment for Jamison to piece together what had happened, but the sound of the orca's labored breathing provided the final piece of the puzzle. The man had shot a whale! Jamison and his wife yelled at the gunman, in disbelief of what they had witnessed. The man simply turned and walked away, ignoring them and never once glancing back at the whale. The two orcas that had passed Jamison immediately turned around, heading toward the animal in distress. As several whales converged, they lifted the injured one to the surface to help it breathe. It became clear to Jamison that the injured whale was a young calf. Despite the efforts of its family members, the breathing of the baby slowed, and the single gunshot wound proved to be fatal. The group slowly dispersed, and the seven-foot-long calf sunk below the surface, never to be seen again.

MOBY DOLL'S IMPACT

In that same summer of 1964, Dr. Murray Newman of the Vancouver Aquarium set out to kill a wild orca. He had hired a sculptor to create an accurate life-size replica based on an orca from the wild for a new exhibit. Newman acquired an exploding harpoon gun from an old whaling station and scouted for a location, ultimately deciding to send his ragtag crew to East Point on Saturna Island, BC. It took six weeks, but the crew finally got their opportunity and fired at a young whale close to shore. The harpoon landed solidly, piercing all the way through the animal's back. When the whale didn't die, Newman ventured out to see it and made the surprising decision to tow the animal back to Vancouver alive on the end of the harpoon line across the Strait of Georgia. The crew was amazed the whale followed them "like a dog on a leash," and its relocation went off without a hitch. They kept it first at the Burrard Dry Docks and later in the harbor itself in a floating net pen.

The public eye was quickly on this newly acquired killer whale. Twenty thousand people came down to see it, and newspapers around the world picked up the story. Scientists from all over the continent came to see the orca firsthand. A pair of marine researchers from a lab in California offered to buy the whale from Newman. He wasn't interested in selling, so he named what he thought was an outrageously high price: twenty-five thousand dollars. Newman was surprised when the men readily agreed to pay the fee, but nonetheless he declined. This whale was not for sale. This whale needed a name, and a member of the media asked if it was male or female. The captors weren't sure but said it was probably a female because of its small size. In the first few days the whale went by several names, but one finally caught on: Moby Doll. Only later did they realize their mistake: Moby was in fact a juvenile male.

Thought to be a member of J-Pod, Moby Doll, with harpoon still attached, in the Burrard Drydocks near Vancouver in 1964. Unknown photographer, Vancouver Public Library, 86934.

During these weeks of up-close observation of a wild orca, old ideas of killer whales being fearsome creatures that devour anything (including people) were shattered. This animal was docile and even curious. Moby was smart too. He made a wide variety of squeaky vocalizations, possibly communicating with whales outside the pen. Most surprising of all, he was gentle. After feeding the whale off a pole for a while, the captors realized they could feed Moby by hand without losing a limb. The orca's health did not improve, however. The animal became sick and listless, finally dying in October. The entire ordeal lasted fewer than three months, but Moby Doll's impact would endure. The widespread publicity, the first-ever positive press for a killer whale, had a variety of profound effects on the lives of wild orcas. With the realization that these blackfish weren't actually dangerous and could be held in captivity, more than thirty aquariums set out to add killer whales to their collections. A lucrative global industry was launched, and every facility that obtained a whale had immediate financial success. Some populations of whales, in particular the Southern Residents, were severely impacted by a decade of live captures.

In addition to the killer whale roundups of the aquarium industry, a different sort of spark was lit among the public. These whales were inherently intriguing, far more interesting than dangerous. Some people were inspired to study them. Moby Doll in particular touched the lives of a couple of young men who went on to have extremely influential careers as killer whale scientists. Michael Bigg, a student interested in harbor seals until he took part in Moby Doll's necropsy, pioneered methods of photo-identifying killer whales and undertook the original surveys of British Columbia orca populations. John Ford, nine years old in 1964, was taken by his mom to see Moby Doll at the Vancouver harbor. Fascinated by the whale, he too went on to become an orca biologist, specializing in acoustics. He started by listening to the recordings that had been made of Moby Doll at the Vancouver Aquarium and used the very same equipment when he was a student to make recordings of other whales the aquarium held. Ford was surprised when, in the late 1970s, he heard vocalizations from wild killer whales over his hydrophones that sounded just like Moby Doll. Through his work we learned that Moby Doll had been a member of the group that came to be known as J-Pod.

THE PENN COVE ROUNDUP

Murray Newman wasn't the only man in the Pacific Northwest interested in catching a killer whale. Ted Griffin, founder of the Seattle Marine Aquarium and destined to become one of the most prolific whale hunters in the United States, also had dreams of keeping a wild orca in captivity in the 1960s. He acquired his first whale, Namu, from a pair of fishermen in British Columbia who had caught the whale in their net in 1965. During attempts to capture his own wild whale, he caught glimpses of how highly social and family-oriented wild killer whales are. Feeling sorry for Namu being without his own kind, Griffin proposed teaming up with his whale-hunting rival, Don Goldsberry, to try and catch a mate for Namu. They partnered to try and harpoon a bull orca from a helicopter, attaching buoys to the dorsal fin via the harpoon so the pod could be more easily followed. The whales became wise to Griffin's tactics, however, and the males eluded him; on more than one occasion, one highly visible porpoising whale had seemingly led the captors away from the rest of the pod, which headed out into more open waters where they couldn't be corralled.

When the opportunity to shoot a female whale with a calf in tow presented itself, Griffin took the shot. The whale turned to dive right as he fired, and as a result the harpoon struck the whale in the side rather than the dorsal area. The adult female died the next day, but her uninjured young calf was brought to Namu and became known as Shamu ("she" + "Namu"). The two animals seemed at first to want nothing to do with each other, with the calf staying as far away from Namu as possible during the first hours in the shared enclosure. Many years later, Namu, from central BC, was determined to be a member of C-Pod of the Northern Resident community. Shamu, captured in Puget Sound, was undoubtedly a member of the Southern Resident community. These two groups don't interact in the wild. Shamu began showing some aggression toward Griffin, especially when he interacted in the water with Namu. He guessed she was too young to be a mate for Namu, which is what he really wanted. With expenses mounting, Griffin made the decision to part with Shamu in hopes of reestablishing his strong relationship with Namu. SeaWorld San Diego agreed to lease the whale, eventually buying her, and Shamu became the face of the franchise for the marine theme park for decades to come (although she lived for only another six years—many other killer whales would use the stage name "Shamu" in the future. Giving all whales the same public stage name helped avoid the media attention around any whales that died, and many did.) Her aggressive tendencies continued at the park, and Shamu was retired from performing after grabbing the leg of a woman riding her as part of a publicity stunt.

Griffin and Goldsberry partnered to capture thirty more whales in Washington State waters as part of ten capture events over the next six years. At least eleven more orcas died during capture, with dozens of others being released. Most whales sold for upward of twenty thousand dollars each, ending up in aquariums around the world. To learn about Griffin and Goldsberry's most famous whale capture, I traveled to the Center for Pacific Northwest Studies in Bellingham, Washington, to see Wallie Funk's photos of a 1970 capture near Whidbey Island that became known as the Penn Cove Roundup. At turns a publisher, an editor, and a reporter, Funk was best known for his journalistic photography in northwestern Washington. One August afternoon, he spent six hours on the walkways surrounding the net pens that contained at least eighty killer whales—likely the entire Southern Resident killer whale population at the time. His 180 photographs of the capture event are somewhat legendary among the regional whale community, having been featured in countless newspaper articles over the years as well as in numerous films and educational displays. The event made an equally large impression on the photographer himself. "I will never forget the experience of those magnificent mammals, indescribable in their grace, movement, and beauty—even in captivity," he later wrote. "The life of a captured orca whose destiny was most likely to be public exhibit in aquariums hundreds, even thousands of miles from its natural habitat and family, touches off at least a pang."

A large body of Funk's work is held on the Western Washington University campus in a branch of the Washington State Archives. The collection is open to public viewing, but research appointments are necessary, with selections made ahead of time regarding which boxes from the collection you wish to view. I wasn't allowed to take anything into the research room but was provided with scrap paper and a pencil after signing in and filling out a form about why I was there. I had selected four boxes that, from the website summary, suggested they might contain some of Funk's killer whale photos. I was instructed to handle the photos with cotton gloves and not to shift the order of the contents in any of the folders. I could mark image sleeves with sticky notes if I wished to pay two dollars an image to have any of them digitally scanned.

I spent an hour viewing the few folders that documented the Penn Cove Roundup, representing just one afternoon of Funk's distinguished newspaper career. I flipped past photos of many notable events on Whidbey Island over the years as well as celebrity photos of The Beatles and The Rolling Stones during their historic visits to Seattle. I bypassed folders labeled "1967 caterpillar epidemic" and "Island Savings Bank Silver Anniversary–1982" to the ones marked "Penn Cove Whale Capture."

I selected nine images to be scanned and gave them to the archivist. Most were pictures of the whales, with nets and ropes around them and boats nearby; one in particular caught my eye because I recognized the whale in it: L26

*L26 Baba with an unidentified young whale at the Penn Cove Roundup.
It's unknown if this whale was her calf or one of the eight taken into captivity.
Photograph by Wallie V. Funk, 1976. Center for Pacific Northwest Studies,
Heritage Resources, Western Washington University, Bellingham, Washington.*

*L26 Baba, forty years after the Penn Cove Roundup, still swam wild and free.
Baba passed away in 2013.*

Baba, her slender dorsal fin with a nick at the top and gaping saddle patch. Beside her surfaces another, smaller whale, just about the right size to have been of interest to the captors. I don't know if this was one of the whales taken or not. In addition to Funk's whale photos, I chose two portraits of the main players—Goldsberry and Griffin, the cowboys in charge of the roundup. Goldsberry has a James Dean look. His wavy dark hair contrasts with his form-fitting white T-shirt, and a day or two's worth of stubble covers his chin. He's looking away from the camera, a cigarette held lightly between two fingers. While Goldsberry exudes masculinity, Griffin comes across as boyish. He wears a heavy knit sweater over a turtleneck shirt. His hair is as long as Goldsberry's but lighter and unkempt, tousled by the wind. He looks straight into the camera with what could almost be characterized as a smirk on his face.

The archivist returned with my thumb drive and brought up the images on a computer in the research room so I could make sure he got the right ones. When he clicked past the photos of Goldsberry and Griffin, he muttered, "Killers." I could tell he immediately regretted the comment. "I'm sorry, that was . . . ." He didn't finish the sentence. Scanning these images had an impact on him, one that reached beyond his occupation and affected him as a human being. "It just seems that whenever someone comes to look at these photos," he explained, "I'm the one who ends up scanning them." I asked the archivist if these photos are viewed by many visitors. He confirmed that they are. "I have a greater affinity for animals than for people," he confided. "It's so hard for me to look at these." Onscreen at that moment was the most graphic of the images I had marked. A dead killer whale calf, bloated and discolored, resting in shallow water. A man wearing waders holds a rope tying the calf's tail to a large cement block. They intended to sink the body. The whales removed from their families for a life of captive display weren't the only casualties from this era.

"LIKE A PRISON CAMP"

The date was August 8, 1970, and a large group of Southern Resident killer whales had just been spotted in Puget Sound. It took only a few minutes for the first speed boat to catch up with them, and before long more boats and several aircraft arrived on scene. The plan was to herd the whales into Whidbey Island's Holmes Harbor using concussive grenades to direct the pod, but the whales had started to wise up to this tactic. This wasn't their first rodeo. Perhaps it was an instinct to protect their young, but I believe the whales had figured out the capture teams were after the smaller orcas. Soon after pursuit began, the whales split into two groups, with mothers and calves diving and heading north while males and other adults stayed near the surface and went east, attempting to act as a decoy. The plan almost worked, as confusion led the captors to pursue the whales they could see, the ones that would be too big to take into captivity. The moms and babies couldn't stay underwater forever, however, and as they made their dash toward Deception Pass (which would lead them to the safety of open waters), they were spotted again by the airborne capture team. The whales were already north of

Holmes Harbor, but the capture team changed directions and rallied to net them off in Penn Cove. The captors were surprised when the rest of the pod, the decoy group, came into the cove soon after. They too were netted up.

Goldsberry and Griffin hired some local youths to help separate the whales. Whenever some of the adult orcas were at the other end of the pen from some of the juveniles, the captors would stretch another net across to partition them off until they isolated the young ones. The whales were frantic, but the capture crews had little concern about setting any of the whales free; if one remained contained, they knew the others wouldn't roam far. The tight bonds between these Southern Residents were evident: the decoy group essentially let itself be caught after the mothers and young had been detained. It took several days to select which whales would be removed, and in the meantime the Penn Cove Roundup became quite the local spectacle. Locals reported hearing the cries of the whales from several miles away. When interviewed decades later, one mother recounted taking her children to see the whales, unprepared for what they encountered. "It was like a prison camp," she recalled. "It was one of the most horrible things I've ever witnessed in my life." She felt her children instinctively knew what they were seeing wasn't right. "Why are they crying?" they repeatedly asked her about the whales. There were thoughts in the community of trying to free the orcas by cutting the nets at night, but with a twenty-four-hour armed guard around the pens, such a plan wasn't likely to succeed.

Eight whales were ultimately selected for captivity, making it the most profitable killer whale capture yet. During the parsing process, one orca mother drowned while trying to reach her calf—the only death that reporters knew about at the time. Unbeknownst to the general public, however, four other juveniles were accidentally killed while sorting the Southern Residents in the net pens. Goldsberry and Griffin covered up these deaths by directing staff to slit the whales' bellies, fill them with rocks, and tie anchors to their tails to sink the evidence. This effort was only partially successful: a few months later, in November, a trawler got the carcasses caught in its nets and deposited them on a local beach, generating bad press that further swayed public opinion against the capture of wild killer whales.

John Crowe, who was eighteen at the time, had been hired to help manage and partition the whales. Looking forward to a paycheck, he had no idea what he had gotten himself into. Crowe remembers clearly the last day of the roundup, when the final whale was maneuvered into a sling for a permanent life in captivity. The team had already started breaking down some of the other net pens, setting the rest of the whales free, but the pods didn't leave. They milled around close to the beach, spyhopping and vocalizing until the last calf was lifted clear of the water. Crowe knew what he was a part of wasn't right, but he didn't know what to do, so he kept working, tears streaming down his cheeks.

LOLITA: THE WHALE THAT SHOULD HAVE BEEN RELEASED

The eight whales that were removed from their families during the Penn Cove Roundup in 1970 were dispersed all over the globe, but only one remains alive today. Tokitae was selected by the veterinarian of the Miami Seaquarium as the whale to add to their exhibit in Florida. He named her after a carving he saw in Seattle with the word "Tokitae" on it, meaning "pretty" in Chinook Jargon, a trade language developed between the Coast Salish and European traders. However, Tokitae was deemed too much of a Native name, tying the whale to her Pacific Northwest home waters, so the aquarium gave her the stage name of Lolita. In Miami she joined another orca, Hugo, who coincidentally was also a Southern Resident. The two performed together for ten years, but after Hugo's death in 1980, Lolita remained alone.

The prospects for Lolita's future changed with the release of the 1993 Warner Brothers movie *Free Willy*, which depicts an orca being captured in the Pacific Northwest but later set free with the help of a young boy who befriended the whale. When the public found out that the whale starring in the movie, Keiko, was in poor health in a marine park in Mexico, a campaign was initiated to make the movie a reality and set Keiko free. Unlike both Lolita and the fictional Willy, Keiko didn't come from the Pacific Northwest. Between 1976 and 1989 more than sixty Icelandic orcas were taken into captivity, and one of these was three-year-old Keiko. After a widespread campaign, significant fundraising, and many twists and turns, Keiko was eventually returned to his home waters in Iceland in 1998. Over the next couple of years Keiko increased his ability to catch live prey and went on dozens of "ocean walks" where his human handlers led him unrestrained into open waters, and during which he occasionally encountered and interacted with wild killer whales. At one point he even swam on his own from Iceland to

Norway, and was feeding himself independently during that time. He always chose to return to his handlers and his pen, however, and showed a continued dependence on people. In December of 2003, Keiko suddenly died.

The story of the world's most famous whale came to an end after a seven-year, twenty-million-dollar rehabilitation effort. Keiko had been the center of controversy for a decade, and questions continued well after his death. Had his attempted release been a success or not? While the Keiko saga unfolded, whale researcher Ken Balcomb turned his attention to Lolita and the prospects for her freedom. The possibility of her release was the best of any captive killer whale. Not only did researchers know more about where she came from, they knew her exact family. In 1995, Balcomb contacted Washington governor Mike Lowry, who got on board with the plan. The men organized a press conference in Seattle, where they announced the launch of the Free Lolita campaign and Lowry declared Lolita a citizen of Washington State. The local media gave the issue some attention, but unlike with the *Free Willy* effort, there wasn't much of a supportive campaign in place to help maintain the initiative. There was no movie to get everyone up in arms, no phone number to call and get involved, no place to send letters of protest or checks of support.

To help maintain the Free Lolita campaign, Balcomb's half-brother, Howard Garrett, helped launch a nonprofit organization called the Tokitae Foundation. Their mission was to secure Lolita's release from her tiny tank in Miami and give her the prospect of a return to freedom, or at the very least a retirement to a net pen in her home waters. Although the nonprofit was initially based in Friday Harbor, Garrett spent two years in Miami leading protests and raising awareness about the orca's plight. Lolita has received bursts of media attention over the years, and the issue of captivity received more attention than ever after the 2013 release of the documentary *Blackfish*, which details the grim realities of life for orcas who live in tanks. Lolita has been in captivity for more than forty-five years. She performs twice a day, every day: that's over thirty thousand shows. She lives in a tank that's barely as deep as she is long, where there is no use for her refined acoustic ability because there's nothing to echolocate and no one to communicate with. She is the only surviving captive Southern Resident of the thirty or so whales that were taken from this population into marine parks during the capture era.

Garrett continues to lobby for Lolita's freedom under the organization Orca Network. One of many lawsuits he's involved with to help Lolita includes an attempt to sue the US Department of Agriculture for issuing permits to the Miami Seaquarium despite numerous violations of the Animal Welfare Act. In particular, Lolita is not provided with any shade from the hot Miami sun, she is a social animal kept solitary, and her small tank does not meet regulation standards for holding a killer whale. In 2015 the National Oceanic and Atmospheric Administration (NOAA) officially accepted Lolita as a member of the endangered Southern Resident community of killer whales, thus potentially giving them a say in her future. If a legal precedent can be set to show that Lolita is being kept in conditions detrimental to her health, Garrett believes this could be used to convince NOAA that she would be better off being retired to a sea-pen in her home waters rather than kept for life in Miami. In 2018 the Lummi Nation joined the campaign for Lolita's release, leading a cross-country trek with a carved orca totem pole to raise awareness and advocate for her freedom. The aquarium continues to maintain that it is in Lolita's best interests to stay in the pen that has been her home for the vast majority of her life.

Lolita, also known as Tokitae, at the Miami Seaquarium in Florida. The only living Southern Resident still in captivity, she has survived in a small tank for nearly fifty years. Photo by Howard Garrett.

Nearly fifty years after the Penn Cove Roundup, it seems that the tide is finally changing on the captive orca industry. After years of campaigning, marine parks like SeaWorld and the Miami Seaquarium have not yielded to moral arguments about cetaceans in captivity, but the movement is starting to affect them in terms they do understand: financial. The captive whales became popular in a different time, when the public knew next to nothing about orcas and had less awareness about the humane treatment of animals. Aquariums have an amazing opportunity to do the right thing and evolve as a business from one that features "Shamu Goes to Hollywood" shows with flashing lights and rock music to one that retires its whales to sea pens where they could truly live out their days as educational ambassadors for their wild counterparts.

THE LAST WASHINGTON STATE CAPTURE

The shift in attitude toward captive whales and dolphins continues to unfold, but these changes began in the 1970s. We were learning more about the intelligence and sociality of orcas through those that had been taken captive, and as a society we started to gain a greater appreciation for whales and the environment in general. Just one example are the recordings of humpback whale songs by Roger Payne that were reaching the public and swaying opinion toward protecting the larger whales who were at risk from commercial whaling. Whales became one of the icons of the emerging environmental movement, as evidenced by the widespread Save the Whales campaign. Greenpeace formed in 1971 and one of their first major campaigns was an antiwhaling effort. In 1972 the US Congress passed the Marine Mammal Protection Act, providing regulation and a precedent to end harassing, injuring, capturing, and killing whales, dolphins, seals, and sea lions in the United States. Permits to capture orcas in US waters were still granted, but public opinion about these events was changing.

In March of 1976, Rich Osborne was a senior at Evergreen State College near Olympia, Washington, as well as an intern at Ted Griffin's Seattle Marine Aquarium. Ever since Osborne had read the book *Mind in the Waters* (particularly the chapter by Paul Spong), he hadn't been able to get whales out of his head. He and his friends hatched an idea to put on an Orca Symposium hosted by Evergreen. A few people were beginning to be recognized as killer whale experts, and the time was ripe to bring them all together in one place. Much to the delight of the students, all the major "orca people" agreed to come and speak at the symposium. A few days before the symposium was slated to begin, Don Goldsberry herded six whales into southern Puget Sound, where they were eventually netted off in Budd Inlet, just a few miles from the Evergreen campus and within sight of the capitol building in downtown Olympia. The capture took place in the public eye, as hundreds of people on boats and from shore witnessed the event. Ralph Munro, aide to Governor Dan Evans, saw it all. He and his wife were in their sailboat and unexpectedly came upon the scene. Explosives the size of beer cans were hurled from a float plane, scaring the whales deeper into Puget Sound and away from open water. Numerous motor boats closely followed the whales, eventually encircling them with nets. Throughout the ordeal, agitated whales splashed at the surface, their eerie cries audible in the air as well as underwater.

The capture was the first successful one in Washington State since the introduction of the Marine Mammal Protection Act. Although it was conducted legally under a permit from the National Marine Fisheries Service, the capture sparked a public outcry. Newspapers all across the country picked up the story. After being filled in on the details of the incident by Munro, Governor Evans immediately called for a review of the permit, pointing out that calls had flooded his office since the event—nearly all of them against the capture. Munro and his wife were so moved by what they saw, he later commented: "It changed our lives."

Protestors flocked to the inlet, many taking to canoes, rowboats, and kayaks with signs demanding the whales be freed. At one point a group of Evergreen students arranged to have flowers dropped on the whales from above as a small plane flew over. With the killer whales showing up practically in Osborne's backyard, he couldn't resist the chance to get closer to them. He convinced a friend to paddle with him in a rowboat out to the whales' net pen. Using a hydrophone he borrowed from the Applied Physics Lab at the University of Washington, Osborne made his first-ever orca acoustic recordings.

The media frenzy increased, and for the most part, coverage took the public's view and were in favor of releasing the whales. Goldsberry responded to the outcry defiantly, boldly defending his actions. "I'm not ashamed," he told the media. "We have done more for killer whales and their welfare than all the environmentalists put together. I'm proud of what I've done." SeaWorld, where the whales were ultimately headed, also came out fighting. "We can almost say these animals will be more healthy mentally in our environment than they would be in the wild where they have day-to-day stresses," said trainer Dave Butcher. "There's no reason for them to want freedom. There's no stress on their stomachs. Not one day will those animals go without food."

The National Marine Fisheries Service insisted the captures had been conducted legally, and Dr. Mark Keyes, a veterinarian for NOAA, defended the captures. "This is going to have some major consequences in the salmon runs of Puget Sound," he said. "One killer whale can eat seventy thousand pounds of salmon a year. At a time when we are extremely concerned about sustaining our fish runs this is a significant factor to be considered." At the time, Keyes had no way of knowing that these whales were in fact marine mammal–eating transients.

One of the whales escaped the enclosure shortly after being caught but stayed near the others outside it. Over the next few days two other whales escaped and departed, leaving three in the enclosure. On March 13, the same day of Evergreen's Orca Symposium, US District Court Judge Morrell Sharp ordered all the whales released—a decision that would later be upheld by the US Circuit Court of Appeals. Sharp questioned whether the permit had been issued legally and believed the capture methods had created such a spectacle in Puget Sound that they were likely illegal as well. He announced that SeaWorld had agreed to transport two of the whales to the Seattle Marine Aquarium for University of Washington researchers. There they would be held temporarily until fitted with radio transmitters that would aid in the study of their swimming and travel patterns after their release.

Osborne wasn't the only college student interested in the whales captured in Budd Inlet. Seventy miles northeast of Evergreen on the University of Washington campus, undergraduate Brad Hanson was also following the story in the media. Born and raised in Seattle, Hanson had been fascinated with orcas since the age of ten, when Ted Griffin brought Namu to the Seattle Marine Aquarium. During that summer, twice a week Hanson accompanied his dad to Pier 66 where he worked, just down the way from Pier 56 where the aquarium was. With fifty cents from his father to pay the aquarium entrance fee, Hanson spent entire days sitting by Namu's pen. Aquarium staff eventually knew Hanson well enough that they let him play with another orca, Kandu, in between shows. He developed a game with Kandu where he would toss the ball and the whale would bring it back. He began to see the intelligence that lay beneath the frightening image of the killer whale, and dreamed of working with and studying them in the future. After high school he enrolled in the University of Washington to get a degree in zoology. At UW he heard about the captured whales in Budd Inlet from a girl in his dormitory. She was getting a group together to protest the captures, but with fond memories of interacting with Namu and Kandu at the Seattle Aquarium, Hanson didn't then see what was wrong with captivity. He declined to protest. Instead, he introduced himself to Al Erickson, the UW professor who had received custody of the two remaining whales. After a month of negative publicity that was starting to affect the University's opinion of Erickson's research, the professor was glad to meet someone so interested. Hanson offered to volunteer if Erickson needed any help, and the professor said he might be in touch.

A short time later, Hanson heard about a one-month position being offered for someone to accompany the two orcas, now named Pender and Flores, to San Juan Island. The researchers needed someone to stay with the whales and assist with tracking them after release. A friend recommended Hanson take a quarter off from school and apply for the job, but Hanson put the idea out of his mind. With no experience, surely he wouldn't have a shot compared with all the graduate students who must be applying. Instead, he picked out courses for the term, but he couldn't get into a single class that he wanted. After returning a second day to the registration tables and again having no luck, he was angry. Storming back to his room, he called Erickson and asked, "Is there any chance you would consider me for the San Juan Island position?"

"Sure," the professor said. "If you're willing to call Don Goldsberry right now and do whatever he says." Hanson agreed and got in touch with Goldsberry, who told Hanson to be at the Seattle Aquarium at 8 a.m. the following morning to help out with the whales. Just like that, Hanson had the job. When he reported for duty, he relieved some very tired workers who had been with the whales 24/7. One of the first people he met in a back room was Rich Osborne.

Erickson planned to install radio tags on the orcas. Citing traditional study methods that manipulated or even destroyed study animals in the name of research, Erickson stood by his belief that the implanted transmitters would be painless to the whales and the information would provide vital insights into the local killer whale population. His colleague, John Sundsten, a professor of neuroanatomy, disagreed with the plan and the two butted heads in the media, their back-and-forth commentary quoted in newspapers. "We ought to treat [the whales] as equals," Sundsten argued, "rather than doing the types of experiments we might do on a rat or even a monkey." Erickson countered: "Few responsible biologists would feel they could get the information needed by watching the animals remotely, especially with killer whales, which are so hard to follow."

With no qualms about the project, Erickson put a radio tag on the female at the Seattle Aquarium (the male's tag would be attached later in the net pen after relocation). The two whales were held in the sea lion tank, and the water was lowered so that the scientists could better access the whale. The radio tags were attached by five pins surgically bored through the leading edge of the dorsal fin. Each tag weighed just over three pounds and was designed to transmit signals detectable from five miles away by boat or three times as far by aircraft. Some members of the media attended part of the procedure, and one reporter wrote about how the whale vocalized every time it was touched with an instrument. Dr. Keyes, the NOAA veterinarian who had spoken out so strongly in the media in favor of the captures, had only one comment. "It's painless," he explained. When the procedure was complete, both whales had radio packs securely attached to them: the #1 pack was on Flores, and the #2 pack on Pender.

In early April of 1976, plans were laid to move Flores and Pender from the Seattle Aquarium to Kanaka Bay on the west side of San Juan Island, where they would eventually be released. Flores would go first, with Pender following a week later. One of Hanson's duties at the aquarium was to attempt to feed the whales by throwing fifty pounds of herring into their enclosure every day. Unbeknownst to Hanson or anyone else at the time, these were transient marine mammal–eating whales, and Flores showed zero interest in the fish. After she departed, however, Pender, always the more curious of the two whales, mouthed and finally ate some of the herring. Flores was the dominant of the two animals, regularly harassing Pender if he didn't swim inside and behind her and once raking him so severely Hanson thought the young male might lose an eye. Hanson suspected that Pender would never have eaten fish had he been with Flores the whole time.

Along with Hanson, Osborne was invited to be on the crew to transport the whales, so he was at the aquarium on the day of Pender's transfer. There was a terrifying moment as the massive whale was lifted out of the tank. While the orca was in the air, the chain attached to the sling snapped, sending Pender crashing back into the water with some of the equipment coming down on top of him. Had the sling already moved the whale from over the tank, Pender would have been dropped on the concrete and severely injured.

Flores and Pender in their net pen in Kanaka Bay before release in 1976. Photo by Shann Weston.

Hanson relocated to San Juan Island along with the whales, staying at Kanaka Bay for the three weeks they were held there in net pens. On the morning of April 26, seven weeks after being captured in Budd Inlet, Flores and Pender were released. Both Osborne and Hanson recall that when released, the whales circled around and seemed disoriented.

They were probably trying to figure out where they were, since they had been transferred abovewater nearly a hundred miles from their point of capture. With funding from the Marine Mammal Commission, Erickson, Hanson, and a few other researchers had use of a boat to track the whales. Flores and Pender were successfully tracked by boat most of the time over the next ten days as they made their way through and around the San Juan and Gulf Islands.

Researchers saw the whales kill and eat several harbor seals and spend one afternoon with another whale known as F1, nicknamed Slash for the huge gash in his dorsal fin. Radio interference caused the scientists to lose the whales, however. Signals were picked up sporadically over the next five months, in part by Hanson camping on top of Mount Constitution on Orcas Island with receivers, but the whales weren't relocated, and finally the signals went silent. It would be three years before the two orcas were seen again. As luck would have it, Osborne was one of the first to encounter them. The two whales were found in Bellingham Bay. The radio packs were gone, but the pins remained in their fins. "It looked to me like the tags had either been scraped off on the rocks or removed by the whales," Osborne recalled. The pins, which would stay embedded in their fins for the remainder of their lives, caused a buildup of tissue that left Pender and Flores with permanent scars.

The ten days of continuous tracking at the beginning of the experiment yielded some interesting data. The pair had traveled an average of seventy-eight miles a day. Their typical travel speed was about three miles per hour, but they reached bursts of over eighteen miles per hour. Their average dives between bouts of breathing lasted about six minutes, but their longest dive was seventeen minutes. Erickson insisted that the radio tags were essential to learn more baseline information about orcas, although other researchers concluded that this type of information could have been easily collected through standard observation techniques rather than via such an invasive procedure. Hanson believes much more could have been done with the data at that time, and he regrets not using it as a foundation for future research.

Hanson eventually went back to school to complete his undergraduate degree at the University of Washington. After graduating, he worked for a time and volunteered at both the Seattle and Vancouver Aquariums. Despite the opportunity to be close to the whales he loved, Hanson realized the structure of the captive setting was not for him. He desired to study wild animals. Hanson returned to the University of Washington for his master's degree with Erickson, tracking and studying the movements of bears in Alaska, and later for a PhD on tag development for small cetaceans. With his education complete, Hanson eventually accepted a job with NOAA, beginning his career as a government scientist.

T13 Flores and T14 Pender became well-known transients in the Salish Sea, easily identified by the scars that remained from their intrusive tags. Flores died around the year 2000, but I got to see Pender on several occasions. He was the first lone bull I ever encountered—a transient who occasionally travels with other groups but for the most part spends his time alone. Some whale-watch captains theorized that Pender routinely circumnavigated Vancouver Island, because he was seen every three weeks or so, always coming east from the Strait of Juan de Fuca and heading north up the Strait of Georgia. That's exactly what he was doing when I first saw him—coming in from Victoria one day and heading north up the Strait of Georgia the next. Pender was last seen in late 2011, and it is presumed he died at an estimated age of forty-three. One of the other whales caught in Budd Inlet but later released, T46, became a very successful mother with numerous surviving offspring; she regularly frequents the Salish Sea with her extended family.

Before the Budd Inlet capture in the spring of 1976, Senator Warren Magnuson of Washington State had urged that Puget Sound become a whale sanctuary. The public outcry after this capture helped Magnuson's bill gain traction and soon after the Senate Commerce Committee voted to amend the Marine Mammal Protection Act to ban the capture of killer whales for display in American waters. The Budd Inlet event was the last orca capture in Washington.

When I saw Pender 33 years after his release, his dorsal fin still bore scars from the radio pack.

LEARN MORE

For more about the live-capture era of killer whales in the United States and Canada, read Sandra Pollard's *Puget Sound Whales for Sale* and *A Puget Sound Orca in Captivity*, Mark Leiren-Young's *The Killer Whale That Changed the World*, Ted Griffin's *Namu: Quest for the Killer Whale*, David Kirby's *Death at SeaWorld: Shamu and the Dark Side of Killer Whales in Captivity*, Erich Hoyt's *Orca: The Whale Called Killer*, and see the documentary *Blackfish*, directed by Gabriela Cowperthwaite.

CHAPTER 7

Whale Organizations:
People, Passion, and Politics

AFTER MEETING KEN BALCOMB IN 1976 at the first Orca Symposium at Evergreen State College, Rich Osborne was keenly interested in helping out with Ken Balcomb's research on wild killer whales, as Balcomb planned to undertake his first orca survey in Washington waters that summer. Osborne, who still had a borrowed hydrophone from the University of Washington, called Balcomb to see if he would be interested in making recordings during his whale surveys that summer. Balcomb said yes, and in August, Osborne became a part of the inaugural Orca Survey team, run under a nonprofit cofounded by Balcomb and two others called the Moclips Cetalogical Society. In the fall, toward the end of the field season, Balcomb returned to another research position and handed over one of his Boston Whaler boats, the *Orca*, and a government credit card to Osborne to finish out September and October. Osborne invited Jim Boran, a fellow Evergreen student, to join the team, and Boran switched his field biology course topic from the northwestern crow to the killer whale.

Ken Balcomb photographs K22 Sekiu as part of the annual Orca Survey for the Center for Whale Research. The Center's research vessels fly yellow flags, indicating a government-issued permit that allows close proximity for high-quality identification photographs.

The original six months of research that became the first annual Orca Survey in 1976 were funded by the National Marine Fisheries Service (NMFS). It was apparent to everyone involved that one field season was not enough to learn about this population of whales, and after having a taste of studying wild killer whales, Balcomb, Osborne, and all the others wanted more. From this initial year of observations it was clear that the whales spent more time in the summer around the San Juan Islands instead of Puget Sound, so the home base for Orca Survey moved to San Juan Island. The problem that followed is similar to one many researchers face today: Who was going to fund the research?

After funding the initial survey, NMFS didn't offer any financial support in 1977. That June, when classes were over, Osborne, Boran, and their friends camped out near Snug Harbor on San Juan Island. All students or recent college graduates, they had spent the winter scheming about how to do more whale research. Balcomb, who had commitments on the East Coast doing oceanic whale surveys, didn't show up until August, so for the first few months, all the whale encounters were shore-based. When Balcomb arrived, he rented what became known as the Orca Survey house on the island's west side for $250 a month. Osborne and crew moved operations to the house and had access to Balcomb's Boston Whalers for boat-based studies. All of them did their research on a volunteer basis. To meet their living expenses, they worked at the Wescott Bay Oyster Farm or as dishwashers at local restaurants; meals often consisted of scavenged roadkill rabbits, homemade baked goods, fish they traded to fishermen in exchange for pies made out of local blackberries, or crabs they caught. Osborne and Boran sold T-shirts and buttons out of the back of their car, offering people the chance to support what was at the time still a novel concept: research of wild killer whales. "It was a labor of love," Boran said years later. "I don't know how we did it."

When Orca Survey arrived on San Juan Island, the locals didn't know that much about the whales. Many islanders didn't even know there were whales around, and those that did—mostly fishermen—were primarily ambivalent. It didn't take long after Orca Survey began its work, however, for the island to become known as the hub for the public's emerging interest in killer whales. Hundreds contacted Balcomb wondering how they could get involved. Although many of the old-timers looked on the newly arrived whale folks as a group of hippies, they were the beginning of a huge paradigm shift: the whales were something interesting to look at—in the wild—in their own right.

In 1978, following another ragtag research season, Balcomb envisioned opening the world's first-ever whale museum in Friday Harbor. The museum could introduce the general public to all the cetaceans of the Pacific Ocean and help fund ongoing research. Balcomb didn't want the museum to endorse or condemn any issues, but rather to be a place to celebrate whales and provide education, allowing people to come to their own conclusions about current issues. It would be a great project to channel all the burgeoning enthusiasm from the public that wanted to help with Orca Survey and get involved with whales. Taking the initial steps to make his dream a reality, Balcomb negotiated a lease for the upstairs of the Odd Fellows Hall on First Street in Friday Harbor. He hired marine biologist Mark Anderson, who had also attended Evergreen's Orca Symposium, as executive director.

Anderson had an ambitious vision for The Whale Museum that included an expert team of advisers, professional exhibits, and a scientific journal published under the auspices of the museum. With Balcomb again on the East Coast, Anderson worked hard throughout the winter in preparation for a grand opening in 1979. About five hundred people volunteered in some way to help create the museum—from work parties to improve the shape of the building to the scientists and artists who contributed to the content of the exhibits. Anderson fondly recalled that time as "one of the most exciting things I've ever been involved in," a sentiment echoed by many of the volunteers who helped turn the vision of a whale museum into a reality. It was a true community effort.

The Whale Museum successfully opened its doors in the summer of 1979, but Anderson and Balcomb's visions for the museum turned out to be vastly different. Unfortunately, controversy fell upon the museum within its first year of operation. The divide resulted in Anderson resigning as executive director in December of 1980 and would plague the whale research community of San Juan Island for the next forty-plus years. Perhaps not surprisingly, there are several different interpretations of what happened. There were disputes about finances, in terms of where money came from and where it was spent, and the rift proved irreparable. Over forty years later, both Balcomb and Anderson expressed bitter regret at having ever entered into a partnership together.

As personnel turnover persisted in these early years, there continued to be a disagreement about whether to focus incoming money on the museum or on research. Under the leadership of a new executive director, the museum board of directors voted to make Balcomb a nonvoting emeritus member, which left him feeling fired from the very organization he helped create. In 1982, fed up with all the politics, Balcomb resigned his emeritus status and cut all ties to the museum, in the process leaving behind some of his early work. Orca Survey, Balcomb's project that had led to the very creation of The Whale Museum, went with him. The museum became an independent nonprofit, and the Moclips Cetalogical Society that had operated both it and Orca Survey became inoperative. After a break from the world of nonprofit politics, Balcomb founded the Center for Whale Research in 1986. While the annual Orca Survey has been run by Balcomb since 1976, in 1986 it became a project of the Center for Whale Research, as it remains today.

Osborne, who was working at The Whale Museum, was now severed by circumstances from his research work at Orca Survey. He wanted to continue studying killer whales, so he worked to establish a research program at the museum. Despite their differences, The Whale Museum and the Center for Whale Research both survived and became regional icons for organizations involved with the Southern Residents, although the lingering politics would never be far from many of their projects. Debate persists today over what type of research and education will be done by both organizations.

THE WHALE MUSEUM: A BRIEF HISTORY

After parting ways with founder Balcomb in 1982, The Whale Museum carried on with several innovative projects. Osborne was named research director and negotiated a lease for the Lime Kiln Lighthouse in 1983. Identified as a prime shore-based whale-watching spot in the 1970s, Lime Kiln was and still is known as one of the world's best sites to see orcas from shore. The lighthouse became the new research station for the museum, and remains so today.

In 1986 the museum's marine mammal stranding network, which had unofficially started in 1981, became federally approved as the official stranding response center in the San Juan Islands. Later, it received grant money to fund necropsies at the University of Washington's Friday Harbor Labs, an important program in learning about the cause of death of local marine mammals. In 1989, after years of negotiations, the museum successfully purchased the Odd Fellows Hall with the help of several grants and a large anonymous donation. This allowed the museum to expand from the second floor to encompass the entire building, doubling exhibit space. The downstairs, where the previous owner had hosted local island events and displayed historic exhibits, became the store and office space. With whale-watching becoming an increasingly popular tourist attraction, the museum took a leading role in guiding the growing industry. Also in 1989, the museum established its early version of guidelines for safely boating around the whales, the first step in what ultimately became Soundwatch, its boater education program.

Several other iconic museum programs have been developed over the years. The Pod Nods program offers educational sleepovers for children, including a flashlight tour of the museum. The Marine Naturalist Training Program features a combination of guest lecturers and field trips on all aspects of the local ecosystem—from geology and intertidal creatures to pinnipeds (seals and sea lions) and, of course, killer whales. More recently, the museum took over hosting a pair of workshops known as Gear Up (in the spring) and Gear Down (in the fall) in Friday Harbor, featuring lectures on local topics to keep working naturalists updated on the latest research. The summer lecture series hosted by the museum plays a similar role but is open to the public.

The SeaSound Remote Sensing Network, founded by the museum in the early 2000s, is a network of hydrophones placed throughout the region allowing researchers and the public to listen to killer whale vocalizations and other marine sounds, from boat traffic to harbor seals and humpback whales. Known today as Orcasound, there are four different nodes in the region, with plans for more. The hydrophones are streamed live on the internet (at orcasound.

net), which has changed the way observers near and far can "watch" whales. People all over the country (and beyond) tune in to listen to killer whales vocalizing live. The network has also made it possible to track some of the whales' movements at night.

THE CENTER FOR WHALE RESEARCH: A BRIEF HISTORY

After leaving The Whale Museum in 1982, Balcomb wanted to see a long-term research program surrounding the Southern Residents continue. The whales were permanently in his heart—the more he observed them, the more he realized it would take a lifetime to begin to understand them. Using the original Orca Survey house on the west side of San Juan Island as a base, Balcomb founded the Center for Whale Research in 1986. The ongoing issue was funding. In the late 1980s a fresh source of money for the center's research became an option. The EarthWatch Institute is a self-described environmental charity, where members of the public pay to have hands-on research experiences with world-class scientists. Since the early 1970s when EarthWatch began, the organization has been involved in hundreds of research projects around the world. The charity is one of the largest private funders of scientific research. Balcomb had worked with EarthWatch in the winter during his work on the East Coast, and through that connection someone suggested setting up a program with the Center for Whale Research. It took some convincing for Balcomb to try it, but he admitted later that it was "the best thing that ever happened" to the center. Groups of ten paying volunteers would come out to San Juan Island for a ten-day period, staying at the Orca Survey house, which was on a large enough property to host the groups. The volunteers stood watch looking for whales, participated in data entry, got to try their hand at photo-identification, and took boat excursions out on Balcomb's research vessels. Amazingly, while running five to nine groups a summer for fifteen years, Balcomb never had a group fail to see whales, although there was one close call. He had to take one group on the boat on their very last morning, then rush to get them to the ferry on time.

The EarthWatch program provided additional benefits to the center beyond just monetary ones. Balcomb is thankful for the structure it provided, as he was required to write up annual reports for the program. As a result, his data is much more organized today, and he recognizes that a lot of information may have slid through the cracks had he not had deadlines to summarize each season's sightings and research efforts. EarthWatch also provided Balcomb with many future staff members, including long-term research associate Dave Ellifrit. Growing up in Missouri, Ellifrit had fallen in love with killer whales as a three-year-old when he saw the movie *Namu*. He didn't see his first whale in person until age seventeen, but as a kid he coveted anything whale-related—from the humpback whale songs he played in his kindergarten class to the whale gift catalog from which he purchased Erich Hoyt's book *Orca: The Whale Called Killer*. "I couldn't believe someone had written a whole book about killer whales," Ellifrit recalled. "When I got home from school one day, there it was, and I was so disappointed because I only had time to look at a few pictures before I had to go to soccer practice!"

Ellifrit learned about the Southern Residents in a 1978 *National Wildlife* magazine article about Ken Balcomb. The photograph of an adult male killer whale with the Seattle skyline in the background made such an impression on Ellifrit that the whale in the picture, J3 Merlin, remained his favorite orca decades later. In 1984 he finally had the chance to see whales when he and his brother signed up for a nine-day excursion with the Oceanic Society to Johnstone Strait in British Columbia. On the way, Ellifrit stopped at the Vancouver Aquarium, where captive Northern Resident Hyak was the first whale he ever saw. In Johnstone Strait he reveled in seeing the A-Pod whales he'd read about years earlier in Hoyt's book. On the way home, as luck would have it, Ellifrit saw Southern Residents from the ferry back to Vancouver, BC, from Vancouver Island. He was hooked.

Ellifrit got his hands on as many whale identification catalogs as he could find. He loved the challenge of trying to identify whales in his or anyone else's photos. In 1988, after a brief stint in the Coast Guard where he learned some valuable seamanship, he returned to the Salish Sea as a student of Rich Osborne's Whale School, an original program at The Whale Museum. That experience convinced Ellifrit that he might actually be able to make a career out of "doing the whale thing." He applied for an internship for the following summer at The Whale Museum. He didn't get it, but serendipitously he came across an EarthWatch brochure that featured a killer whale program. When Ellifrit realized the opportunity was with Balcomb, one of his heroes since that 1978 magazine article, he knew he had to go.

Ellifrit was several steps ahead of the game when he arrived at the Center for Whale Research in 1989, as he already knew most of the whales from poring over identification guides back home. He doesn't remember seeing much of Balcomb during that initial EarthWatch stint, but his ID skills made an impression on Balcomb's wife, who must have passed the word on. On Ellifrit's last day, Balcomb drove him to the airport and casually asked if he would be interested in coming back the next summer. Ellifrit didn't have to think twice about his answer. Two days before Christmas later that year, he received the best piece of mail he ever got: a letter with the Orca Survey annual summary, with two handwritten sentences at the bottom from Balcomb formally inviting him back as staff for the summer of 1990. Although Ellifrit wasn't paid, he was given food and a place to stay. He recalled having no trouble living out of a sleeping bag since it meant living his dream.

Ellifrit's arrival at the center was timely. In October of 1990, Michael Bigg, the father of modern killer whale research, lost his battle with leukemia. For years, Balcomb had sent all his photos from each season to Bigg and colleagues for official identifications. Despite being the ace photographer in charge of a long-term photo-ID study, Balcomb admits photo-identifications were never his strong suit. Balcomb was impressed by Ellifrit's knowledge of Southern Residents as well as Northern Residents and transients. Ellifrit easily slipped into the role of photo-ID expert for the Center for Whale Research. His identification skills come not from a photographic memory but from countless hours spent looking at and comparing pictures. Once he knows a whale, he has a great ability to retain the information, but he's so good because of his dedication to the task.

Indeed, Ellifrit's talent goes well beyond looking at just dorsal fins and saddle patches. He has an innate sense for the gestalt of each particular whale. "J26 looks a lot like J3 [his probable uncle]," Ellifrit explained to me one evening. "Not in the saddle patches, but parts of them just hang the same. They both have the same flat back." While he's seen "every rule broken" as far as morphological trends across different ecotypes, he thinks there are subtle morphological characteristics genetically passed on within given matrilines. The transient family the T65As all have the same smushed-looking faces, for example, and the K12s all have nearly identical eyepatches. Ellifrit's eye for detail has important implications for long-term monitoring of this population. Some whales just appear skinnier than others, for instance, so you can't say a whale looks malnourished unless you know its personal history. "If I saw L92 out of nowhere, I'd say that whale was about to die," Ellifrit explained, "but because I've watched him I know he's had that peanut-head since he was a juvenile." After decades of experience, Ellifrit can tell when a female might be pregnant based on nuances in how she's surfacing. The smaller females, when pregnant, seem to have a stiffer back. It's an astonishing skill, honed by decades of practice. It's no wonder Ellifrit found a permanent home at the Center for Whale Research.

There has been an entire other cast of characters at the Center for Whale Research over the years, people who have returned to help Balcomb with the Orca Survey efforts and EarthWatch groups, most all of them doing so on a volunteer basis. The 1990s were the center's heyday, as the Orca

Survey house was a hub of activity every year from May through October. In addition to a full house, numerous people set up camp in the yard. For the staff, every summer was a reunion with the whales and each other. Every night was a party. "The nineties were my sixties," Ellifrit reminisced.

Around the year 2000, however, things began to change. EarthWatch decided that licensed captains had to take the paying volunteers out, which meant the center could no longer take EarthWatchers out on their own research vessels. That changed the experience significantly, although they continued for several years contracting a local commercial whale-watch company. Shortly thereafter Balcomb was diagnosed with cancer. The EarthWatch program at the center concluded in 2004, just as the Southern Residents were being considered for listing on the Endangered Species List. While Balcomb won his battle with cancer, the center maintained a lower profile after that, functioning with a much smaller research team. Many of the regulars became scarcer, although Ellifrit remains a constant presence.

The Whale Museum and the Center for Whale Research are among the longest tenured whale organizations in the Salish Sea, but many other groups, institutions, and non-profits have formed over the years with orcas, and particularly the Southern Residents, at the core of their mission. Orca Network spearheads the Free Lolita campaign and has become the central sightings network for wild whales in the Salish Sea (see chapter 6 for the details of Lolita's story). They also participate or host educational events and workshops throughout the region, including operating the Langley Whale Center on Whidbey Island. Killer Whale Tales provides environmental education through storytelling, bringing real-world research experiences into elementary school classrooms and introducing children to the orcas and their plight. Whale Scout puts naturalists on the shorelines to educate whale-watchers while sharing real-time sightings; the organization takes a boots-on-the-ground approach to get whale advocates involved in local stream restoration efforts to help with salmon recovery. The Orca Behavior Institute is the independent citizen science research nonprofit I cofounded to focus on noninvasive behavioral and acoustic research on regional killer whales. These are just a sampling of the regional groups that study, educate, and advocate for the orcas of the Salish Sea.

NAMING THE WHALES

Ever since primatologist Jane Goodall began naming the chimpanzees in Tanzania's Gombe Stream National Park in the 1960s, the practice of naming the animals we study has become more prevalent. Although highly controversial in the early years for breaching protocols that were meant to keep the observer separate from and unbiased toward the subject, Goodall's methodology achieved a different goal: it helped people relate to and establish connections with another species, one they may have never seen in real life. Bigg and the other early researchers studying the Southern Residents in the 1970s learned that, like Jane's chimpanzees, orcas could be individually distinguished. Instead of having unique faces like primates, orcas have distinctive dorsal fins and saddle patch markings that allow humans to tell them apart. Because of these markings, researchers started to piece together the picture of the regional killer whale population: they learned that the same whales returned to the area day after day and year after year and that some of the same whales were nearly always in association with one another in tight-knit family groups.

It now seems natural that researchers developed nicknames for the whales they recognized and spent a lot of time with. J1 Ruffles was named for his wavy dorsal fin, and K40 Raggedy for the tattered trailing edge of her dorsal fin. J2 Granny, J4 Mama, and J12 Sissy were named for their perceived familial relationships at the time. Other whales, like J7 Sucia, J8 Spieden, and J18 Everett were named after local islands or other geographical features. Some orca names, such as J14 Samish and K7 Lummi, paid tribute to local Native tribes, while others were nicknamed after pop culture characters of the era, such as J21 E.T. and K20 Spock. One way or another, by the early 1980s researchers had developed common names for all of J- and K-Pods in addition to their official scientific alphanumeric designations.

The first official effort to name the Southern Residents occurred in 1983 when The Whale Museum and Greenpeace Victoria launched a campaign to stop Sealand of the Pacific from capturing two young L-Pod whales, for which they had received a permit from the Canadian government. Plans were in place to trap the whales in Pedder Bay, near Victoria, but before the captures would have a chance to take place, the goal was to establish a connection between the public and the wild whales. With an article in the *Victoria Times Colonist*, the first name-the-whales contest began, with readers asked to submit suggestions for all of the fifty L-Pod whales. The entry form was headed by bold lettering asking readers to "Help Keep L-Pod Together" by participating in the contest and pledging their support to the cause. The article acknowledged that there were concerns about being gimmicky or "taking a 'show biz' approach to animals in the wild by giving whales personalities similar to those in captivity." Ultimately, those behind the naming campaign felt the educational benefits of the contest far outweighed any disadvantages. If the public could connect to these whales as individuals before the captures were to occur, they might be prevented from occurring at all. Whale researchers selected winners from the entries, and all of L-Pod was officially named. The campaign was successful and Sealand captured no Southern Residents, although instead they got their orcas from Iceland. Naming the whales helped their popularity grow among the public, and in early 1984 with a congressional bill pending to ban all live killer whale captures in the United States, The Whale Museum launched the Orca Adoption Program, using the researcher-derived nicknames for J- and K-Pods and the contest-winning names for L-Pod.

In 1988 researchers discovered that they had misidentified the gender of some of the whales and did some "gender adjustments". Adult males and females can be distinguished by their overall size, particularly the height of their dorsal fin, but juvenile whales can only be identified as a male or female by the black and white markings on their belly. It's not easy to see the underside of a whale and simultaneously be confident of which whale it is, since you need to see the saddle patch on the back to be sure. Today, there are many more eyes (and cameras) on the water, so Southern Resident genders are usually determined with more speed and accuracy, but in the earlier decades, some whales didn't have confirmed genders until adulthood. Three whales whose names were not gender appropriate actually had their names changed: K14 Leon became Lea, L14 Cordelia became Cordy, and L44 Cleo became Leo. Other whales—like males L41 Mega, L57 Faith, and L61 Astral—had their genders "adjusted" from female but kept their names. It was decided that gender-neutral names should probably be chosen in case this happened in the future. Researchers now knew to be more careful when designating genders, but a few errors still happened, even in later years. J16 Slick and K20 Spock were both considered males and had even been listed as having the "fin sprouts" characteristic of teenage males when they returned with their first calves, thus confirming that they were in fact females.

Throughout the 1990s calves were named at the end of each summer at a party that included members of both the Center for Whale Research and The Whale Museum. Orca adopters were encouraged to submit ideas, and researchers also made suggestions. Staff members from both organizations picked the final names. This kept a large part of the whale community directly involved with naming the whales, and the ideas for whale names were similar to those used in the early years. Many were named after local geographic features such as L72 Racer, who was first seen near Race Rocks. L79 Skana was named after the Haida word for "orca." Pop culture icons also still played a role; L95 Nigel is the result of a summer where the whale researchers were into the mockumentary *Spinal Tap*. Other whales were named after favorite dogs that had passed away, such as J31 Tsuchi.

On some occasions, whale names were a form of paying tribute to someone. Although there has always been an unofficial policy not to name orcas after a specific person, there have been a few exceptions, such as J6 Ralph, so named for Ralph Munro, who helped stop orca captures in Washington and also helped to designate Lime Kiln as a state park. Another exception is J26 Mike, named for Michael Bigg, who passed away the year J26 was born. A few other whales were obliquely named in reference to people. K27 Deadhead was named to honor the passing of Jerry Garcia of the Grateful Dead. However, her name has a double meaning because "deadhead" is also a nautical term referring to a floating log. In 2012, Deadhead's first calf was named Ripple. This name seems like another water reference, but it's also a the title of a Grateful Dead song. Another established naming practice is that the descendants of J14 Samish are named by the Samish Nation during traditional ceremonies. This practice began in 2001 with J37 Hy'Shqa, named after a term meaning "blessing" or "thank you." J40 Suttles was named in honor of Coast Salish researcher Dr. Wayne Suttles; J45 was named Se-Yi'-Chn, meaning "younger sibling"; and J49 was named T'ilem I'nges, meaning "singing grandchild."

The fact is that the whales don't care what they are named—only we humans do. But that doesn't make their names any less important. As a marine naturalist, I learned that one of the easiest ways to establish a connection between human visitors to the Salish Sea and the Southern Residents is to share their names. If you're out on a whale-watching excursion and you point to a whale and say, "That's J2 Granny. She was born in these waters and is almost a hundred years old," or, "That's J28 Polaris with the notch in her fin. She is always with her mom and her sister," people remember it. When this knowledge is right in front of them, they latch onto it and recognize when Polaris surfaces again. They might ask if the whale next to her is a relative. Suddenly, they've made the connection that these animals aren't just generic whales; they are individuals with personalities and families just like people.

An idea promoted by The Whale Museum was to let orca adopters fully decide the names of the whales. Suggesting names would be a privilege reserved for adopters, and allowing the public to vote on names would ensure that chosen names would appeal to a wider audience. J34 DoubleStuf and L91 Muncher might be "adoptable" names, but locally they were unpopular and deemed silly. The whale-naming issue thus started to get political. While the method of having the public name the whales made sense financially for The Whale Museum, the result was that many of the people who devote their lives to studying the whales and educating people about them were now left out of the naming process. The Center for Whale Research, who gets no money from the Orca Adoption Program but is in charge of the official census of the orca population, went from hosting whale-naming parties to having no say whatsoever.

The issue came to a head in 2010, when Ken Balcomb wrote a blog post on the website of the Center for Whale Research naming seven of the whales, including two that had already been named by The Whale Museum and three that were in the process of being named that very month during the museum's annual contest. For the first time, some of the whales had received conflicting names. The Whale Museum called L110 Midnight. The Center for Whale Research would now be calling L110 Flapper, for the flap of skin that he bears as a scar along the right side of his mouth, the result of an injury when he was just two months old. Suddenly naturalists, formerly neutral parties in The Whale Museum/Center for Whale Research divide, were forced to pick sides over something as inane as whale names. From an outside standpoint it seems comical, but those directly involved in the controversy took it very seriously. The naming issue raised some challenging questions about rights to and ownership of these whales. Surely Balcomb, involved in studying the Southern Residents from the very beginning in 1976, deserved to be involved in the naming process. Yet so did the orca adopters, those who must establish and spread a connection to this endangered group of whales if they are to survive. There must be some middle ground, but the contentious whale community has yet to find it, on this and many other issues.

SOUNDWATCH AND THE PACIFIC WHALE WATCH ASSOCIATION

In the summer of 1994, the School for Field Studies and The Whale Museum promoted Rich Osborne's idea for two No Sound in the Sound Days, to raise awareness about the potential impact of vessel noise on whales and to experiment with alternative management practices that might include no-boat zones. The efforts, endorsed by more than eighty businesses and organizations (including the commercial whale-watch community), proved how varied interests can indeed come together when it comes to wanting to protect the Southern

Residents. The collaboration of the whale-watch companies on this project influenced their decision to form their
own organization: Whale Watch Operators Association
Northwest (with the clunky acronym WWOANW, later
renamed the Pacific Whale Watch Association, PWWA).
The whale-watch association started to hold its own annual
meeting to train drivers and discuss changes to their existing
voluntary guidelines for boating around the whales; these
meetings eventually took the place of the annual community meetings that had been hosted by The Whale Museum.
The School for Field Studies launched what would become
the museum's Soundwatch Boater Education program, an
on-the-water education platform about safe boating practices
around the whales and a monitor for how well the PWWA
was following their own guidelines. Soundwatch and the
whale-watch association would work together to manage
vessels around the orcas for years to come.

PWWA members have always discussed how to best
operate boats around the whales. They have made use of
adaptive management practices, where their voluntary
guidelines are continually updated based on the latest
information and the best available science. Although it
hasn't always been easy to keep everyone from dozens of
different companies on the same page, the organization has
lasted because they all share the same common denomina-
tor: they see themselves as responsible capitalists, and they
want to do what's right for both their businesses and the
whales. The association has been plagued by politics over
the years, but they've stuck together because it's made sense
for them do so, particularly in response to an opposition
that believes whale-watching is part of the problem in the

decline of the Southern Residents. It's not always an easy balance. PWWA members want to run a good business and make sure customers leave happy, but other groups, such as the Orca Relief Citizen's Alliance, claim the whale-watch industry is loving the whales to death. Despite some opposition, the PWWA has served as a model program for other whale-watch communities around the world.

Two whale-watch boats out of Vancouver, BC, observe J2 Granny in Cattle Pass near San Juan Island.

Kari Koski was hired as the first director of the Soundwatch Boater Education Program. She grew up outside of Juneau, Alaska, where as a child she taught pretend natural history classes to her imaginary friends. Fascinated by whales from an early age, she thought she wanted to be either a dolphin trainer or a whale researcher when she grew up. When Koski was twelve, a family friend took her to SeaWorld for the first time, where she hated the killer whale show and thought the petting pool with bottlenose dolphins and false killer whales made people crazy as they splashed and tried to lure the animals closer. A year later, in 1984, her father's department at NOAA was involved in denying a SeaWorld permit to capture one hundred killer whales in Alaska. "I remember being horrified by the thought that the whales we saw at home could end up in a place like [SeaWorld]," Koski recalled. "That, along with a whale-watch trip we took in Hawaii, convinced me I was not going to be a dolphin trainer."

When Koski got accepted to Oregon State University, she initially thought she'd become a whale biologist but was turned off by what she viewed as a male-dominated field full of big egos. "Also, the research wasn't people-oriented enough," she said. "It wasn't connected to social issues, it was really competitive, and it seemed like everyone was in it for themselves." She loved biology but thought education might be more her route. In the end, Koski graduated with a degree in general science but feels like her college curriculum was essentially an environmental science program. After college, she was very aware of the impending crisis between tourism and whales. She wanted to do something about it, and when Osborne told her about the School for Field Studies program that became the pilot version of Soundwatch, Koski felt like her future was handed to her on a silver platter.

In her role as Soundwatch director, Koski saw herself as the conscience of the whale-watch business. She viewed most humans as "takers" from the environment, and she was driven by a passionate belief that things could be done differently. As a community, for example, whale-watchers could give back to the whales, and indeed their conservation programs have grown to the point that they are donating hundreds of thousands of dollars each year to regional non-profits. Soundwatch and the whale-watch industry were true partners for many years. The strong communication between the two entities allowed whale-watch guidelines to be modified each year based on feedback from both sides. Together, Soundwatch and the commercial whale-watch community created a voluntary no-go zone for boats around Lime Kiln and San Juan Island's west side when whales are present. Other modifications have included the guideline to approach whales at slow speeds and buffer zones along shorelines. Discussions about rule changes have led to some hard decisions that were trade-offs for the companies, but they were based on Soundwatch data and agreed to by everyone. The industry described Koski and Soundwatch as the "watchdog" of their business interests. They even turned to her to help train their new drivers.

In 2001, Canada's Department of Fisheries and Oceans partnered with the Veins of Life Watershed Society to start a sister program to Soundwatch called Marine Mammal Monitoring (M3). Later on, another Canadian program called Straitwatch played a similar role. In 2002 the US and Canadian governments, the whale-watch industry, Soundwatch, and M3 all collaborated to create a uniform and concise set of voluntary whale-watch guidelines called Be Whale Wise. Part of this project was the creation of a brochure, paid for by both governments, that could be handed out on shore and on the water to private boaters to educate them about how to operate their vessel safely around the whales. In addition to these more straightforward guidelines, the commercial whale-watch association continued to follow their own, more detailed rules agreed upon and partially self-policed by their own drivers. Starting in 1997,

Soundwatch sent out feedback reports to each company noting any violations they had seen from their boats during their hours monitoring on the water. This allowed the owners and operators of the company to get some feedback on how well they were following the agreed-upon rules without any component of "public shaming," although there was definitely on-the-water peer pressure from other operators to hold to the agreed-upon standards.

A typical day for Soundwatch drivers and volunteers during the summer involves meeting in the morning at Snug Harbor on San Juan Island, where the Soundwatch boat is moored. When they hear of a whale report, everyone gears up in warm, waterproof survival suits and heads out. While out with whales, Soundwatch tends to stay either ahead or behind the group so they can intercept private boaters heading toward the animals. This way, they can stop them and hand over the whale-watch guidelines before the boaters reach the whales. In addition to this on-the-water education, Soundwatch conducts hourly surveys of all boat traffic within a certain distance of the whales and notes any infractions seen by either commercial or private boaters. In the small, open-air boat with no shelter or bathroom facilities, it can be a long day out there—sometimes up to ten hours.

For a couple summers I volunteered on Soundwatch while interning for The Whale Museum. Dealing with the boat and the elements was one thing: being out in the sun all day; eating snacks out of ziplock bags when you got a break between writing data or handing out brochure packets; handling the pounding of the boat over the waves when running at high speed, which sometimes led to getting

The Soundwatch Boater Education Program, run by The Whale Museum, hands out a copy of the Be Whale Wise guidelines to a private boater in Haro Strait.

drenched with saltwater; and of course, peeing in a bucket if you couldn't hold it any longer, hoping that most people on the dozens of boats in front of you were looking the other way. But all that was just part of it. The other part was interacting with people on the water, which was sometimes an adventure in itself.

For the most part, people were understanding and interested. They usually slowed down or stopped their boat when we hailed them, and while many of them were ignorant of the fact that there were whales in the area, they took the guidelines and listened to our instructions about how to best get a look at the whales or transit around them. Some people were polite, then more or less did whatever they wanted, whether that was getting too close to the whales or operating their boat too fast in the whales' vicinity. In those cases, there was nothing we could do but write them up in our incident report; if the education wasn't effective, Soundwatch's hands were pretty much tied—contrary to what some people believed, Soundwatch can monitor boat traffic but has never had any enforcement capabilities. When I was out with Soundwatch, mostly between 2003 and 2005, all the whale-watch rules were just guidelines—there weren't any regulations that could be enforced in the United States outside of the Marine Mammal Protection Act, which would take a pretty egregious act of harassment for a boat driver to be successfully prosecuted.

The absolute worst incidents were the people who flat out didn't care about the whales at all. These boaters wouldn't slow down to talk to us, maintaining their fast speed and looking in the other direction or, even worse, flipping us off or cursing at us as they drove by. When they were willing to engage, they were arrogant and rude. I'll always remember one man who, after hearing about the whales and what distance he should stay away, responded, "Yeah, I really don't care about any of that stuff." He then proceeded to transit—at high speed—right over the top of the whales. While encounters with people like that were the most memorable, they were in the minority. Most people didn't know what was going on, but once they received education from Soundwatch, they wanted to do the right thing.

The partnership between the commercial whale-watch industry and Soundwatch continued to function well until the Southern Residents were proposed for endangered listing, first in Canada and then in the United States. With vessel effects on the list of potential risk factors to the whales, the whale-watch industry had to take up a more defensive position. The rift with Soundwatch formed when the program shared data on vessel infractions around the whales with NOAA; the understanding of the PWWA was that this was proprietary information only to be shared between Soundwatch and the association. They felt their trust had been breached and the data was now being used against them. When the endangered listing passed in the United States in 2005, Soundwatch received its first government funding to help NOAA monitor the potential vessel issues. Financial support for Soundwatch from the PWWA stopped at about this same time. With Koski obligated under government contracts to share boat-monitoring data to the managers who would be creating formalized vessel regulations, she started to be viewed more as a policewoman rather than a partner or ally.

In 2009, Washington's Department of Fish and Wildlife announced a law prohibiting vessels from approaching killer whales closer than two hundred yards in Washington waters, doubling the voluntary guideline distance of one hundred yards that all whale-watch boats had been following. This same year, NOAA initiated a process of considering alternatives for various vessel regulations, a process that proved controversial and divisive as it unfolded over two years. After advocates on both sides of the issue strongly voiced their opinions both in public meetings and via submitted comments, NOAA announced their federal regulations in 2011. These included making the two-hundred-yard rule a law at the federal level but did not implement the more controversial no-go zone for boats along the west side of San Juan Island. On-the-water altercations involving commercial whale-watch drivers and Soundwatch increased. The industry stopped inviting

Koski and Soundwatch to their annual meetings. With new Washington State vessel regulations in place and enforcement boats on the water for the first time, Koski's worst year on the water was 2009. While it wasn't apparent if or how the updated boater regulations would be enforced, Koski recalls, "It was clear to me that we were no longer partners at that point." By the end of 2011, she had taken more than she could stand. Koski retired from Soundwatch in early 2012.

Soundwatch has continued under The Whale Museum, but the program has had a turnover in directors and challenges with funding. They're not on the water as much, and when they are, they are seen by many as being less effective than they used to be. One longtime whale-watch captain stated, "Soundwatch has become somewhat nonfunctional. I don't know how [Koski] put up with everything for as long as she did, but the program has gone downhill without her. She was so good. She had a lot of respect." Another captain said: "I'm afraid for that program. Even when we disagreed, we could still sit down and talk to [Koski] because there was a long history there, but after she left, the continuity of the program fell apart. There's still a value to having them out there, and their role is an important one, but they aren't executing it very well right now."

The endangered listing of the whales and subsequent vessel regulations didn't only fracture the partnership between the PWWA and Soundwatch; it put enforcement boats capable of handing out fines on the water for the first time. San Juan County and Washington State's Department of Fish and Wildlife enforced the first official vessel regulations, with NOAA providing additional enforcement after the federal ruling went into effect. Many critics of whale-watching were happy to finally see an enforcement presence, but as with seemingly all things surrounding the whales, it was not a simple issue. The main problem was that the operators of the enforcement boats had much less experience boating around whales than the commercial whale-watch captains. It has been difficult to get the enforcement drivers up to speed on the whale-watching scene, and really, they need the time and experience on the water that the commercial operators have.

One man who does have that experience is Brian Goodremont, owner of San Juan Outfitters and San Juan Safaris. Goodremont came to the San Juan Islands by way of Ohio in the late 1990s as part of a summer college course. He fell in love with the place, and the whales, and returned the following summer to work as a naturalist. After several years he got his captain's license and started managing a local whale-watch company before beginning his own business. He served as president for the PWWA for three years and remained involved in the executive management of the association for many years after that. I met with Goodremont to hear his perspective on the vessel regulations. He has a lot of respect for the difficulty of the task set before the enforcement officers, and he appreciates that they do use their discretion on the water. It's impossible for anyone to obey the letter of the law 100 percent of the time when observing the unpredictable killer whales (who of course don't know or follow the regulations themselves), and if enforcement wrote up every infraction, there would be hundreds if not thousands of incidents every season. However, in their plus or minus thirty days on the water each year in 2013 and 2014, Washington Department of Fish and Wildlife only handed out two or three citations. In

2015 that number increased slightly but not substantially. Goodremont thinks the enforcement teams still have some way to go. He recalls one incident where the enforcement boat motored right over the top of killer whales to talk to a parked sailboat. This is a classic example of the enforcement boats causing more potential vessel disturbance to the whales than was there without them. I have several similar observations of my own.

The internal voluntary guidelines followed by the PWWA, Goodremont points out, have always gone above and beyond what the regulations or Be Whale Wise guidelines ask for. However, he has major issue with the wording of the federal regulations that require boats to "maintain a distance of two hundred yards" at all times, meaning boats have to constantly try to move out of the whales' path if they change directions. "NOAA has always told us that the potential bad things for whales from vessels are unpredictability, moving propellers, engine exhaust, and noise. Needing to maintain a distance of two hundred yards means that we have to do more of all of those things." He describes a scenario where the whales are foraging, and hence moving somewhat unpredictably. The ideal thing to do, he thinks, would be to shut the engine down and sit quietly while the whales go about their business milling this way and that. Goodremont would also be able to drop his hydrophone in this scenario, adding to the experience of his guests, and if other boats were able to stay stationary, there would be little vessel engine noise. The way the law is written, though, he needs to keep his engines running and try to maneuver out of the whales' path every time they change direction. "It just doesn't make sense," he concludes.

From my own years of observing the whales, the boats, and listening to acoustic recordings, I agree. The largest potential impact boats have on the whales comes from the noise underwater. I've listened to recordings where I had to pull the headphones off because the boat noise was so loud and constant that I couldn't take it anymore. Whales don't have that option. The idea that boats should start their engines to back out of a whale's path if it changes directions seems counterintuitive to me; the best thing to do is stay a predictable, parked, quiet floating object and let the whales just go about their business.

The PWWA banded together not only to come up with standardized rules for whale-watching but also to share sightings. One Canadian member of the PWWA recounted how the first two whale-watch companies out of Victoria tried in the early years to keep sightings from one another. When they continually ran into each other early in the morning scouting for whales from the same shoreline loca-tions, they realized it would make more sense for them to pool their resources. It's a difficult balance, one that has plagued the whale-watch association from day one: they're competitors in business, but in many ways it makes sense for them to work together.

The association's sightings network has taken many forms over the years. Early on, reports were shared pri-marily over marine radio. The industry developed a code, communicating whale identities and locations in ways only their members would understand. At times, certain opera-tors might find whales and not share the report right away with their colleagues. This led to infighting, where certain factions of companies might not want to share reports with another company that had previously withheld information from them. In the early 2000s the companies started using a pager system, sending out coordinates, whale identity, and direction of travel in a code system that might look something like 905-30-80-1-1400, meaning J- and K-Pods (905) were at Kellett Bluff (coordinates 30 and 80) heading north (1) at 2:00 p.m. (1400). Much of this data came from a paid PWWA member with "big eyes" (large binoculars) sitting atop a hill near Victoria with a vantage point overlooking much of the Salish Sea. As technology advanced and cell phones became more prevalent, the pagers and deciphering the code became more cumbersome and eventually obsolete.

In the mid-2000s, Ron Bates, a native Vancouver Islander and one of the region's original whale-watchers, centralized sightings from the most peripheral whale-watch companies in Vancouver, Sooke, Port Townsend, and Anacortes. If they called him with whale reports, he would quickly dispense information out to everyone. This proved effective, and with Bates well respected throughout the entire whale-watch community, he became "whale central" for the PWWA. His only compensation was having his cell phone and radio paid for (as well as free coffee and the occasional beer). Bates tirelessly and unfailingly received and dispensed whale reports on a daily basis for a decade. The primary mode of communication was text message, where detailed whale information could be dispensed directly and privately to owners and drivers without the need for code. The manager of Five Star Whale Watching out of Victoria laughed about working at the same company as Bates: "People always tell me, 'You're so lucky to work at Five Star, you probably get whale information first all the time.' In reality, we're usually the last to know. Ron is so focused on there being a level playing field that the text message is his first priority. He'll be focused on his phone typing away as we're pulling out of the harbor, and meanwhile our driver will be saying, 'Ron? Ron? Which way do I need to go?'" A private Facebook sightings page is the most recent evolution of the sightings network, with all member companies able to report and receive sighting information quickly and easily without one person being responsible for sending out current reports.

Despite some conflicts between companies over the years, the relationship among drivers on the water has truly seemed like a brotherhood. (Although a few women have joined the ranks of whale-watch captains over the years, the community is predominantly men.) While sightings information is mostly communicated "off the air," a lot of chitchat occurs over the public airwaves of VHF radio, and the sense of camaraderie has always been apparent. There's a certain lingo shared among whale-watch captains, with special words and phrases that are half code to keep specifics hidden, half slang developed by the niche industry.

A DARK SIDE

The sad truth is that there is a dark side to the whale community in the Salish Sea. On the one hand, amazing people have come together with passion and dedication to watch, study, and educate others about the whales. But on the other hand, the longer you are part of the whale community, the more entrenched in human politics you become.

In so many of the interviews I did for this book, people spent at least a few minutes talking about one unfortunate phenomenon: these whales seem to bring out the worst in us. It's not that the whale community isn't made up of good people. I firmly believe everyone wants what's best for the whales. But the bizarre, divisive history of the regional whale scene means that you pick sides without even knowing it. To stay involved means you start doing things that uphold the boundaries we humans have placed among ourselves: you're either a boat-based whale-watcher or a shore-based whale-watcher, you're with our company/ nonprofit or theirs, you're with us or against us when it comes to issues like commercial whale-watching, invasive research methods, or the best routes to recovery. There's also a bizarre competitive undercurrent to whale-watching, where everyone wants to be the first on scene, the first to see a new calf, the first to determine a baby's gender, the first to post photos of an amazing encounter, to see the most and the best and miss nothing.

Many people I interviewed expressed stories of the wrongs that have been done to them by other members of the whale community. Similarly, these very folks have also done things they are not proud of, in the name of maintaining their position in this community. The list of reported wrongs is deep. Boxes of historic photographs or audio recordings have been stolen from their rightful owners. Data has been published under the names of people who didn't collect it. Accounts of sexual favors being performed for continued access to boat trips with the whales have been reported. Good science has failed to get published because somebody wouldn't sleep with somebody else. Videos have been posted on social media to publicly shame former friends. Tens of thousands of dollars have been misspent or conveniently misdirected to improper projects. People have been forced into resignations for personal vendettas. False information has been fed to the media, or presented in order to secure funding over somebody else. Celebrities have backed out of lending support because of behind-the-scenes phone calls. Even the publication of this book was jeopardized on more than one occasion because of politics. There's discrimination—based on gender, or past affiliations, or personal relationships. Friends have backstabbed each other, and former allies have gotten into public yelling matches. Some people keep their data literally under lock and key, they're so afraid it will be stolen. Certain parties won't invite others to the table to discuss an issue, because they're wary of what will end up in the media, or that what they reveal about themselves will be used against them, or because those two groups just don't talk to each other anymore. In general, there is very little willingness to collaborate. You can only stay if you follow the rules—the rules of us versus them—so tow our party line or be ostracized.

Many in this community have lost faith in those they used to look up to. One person told me: "It's possibly the biggest life lesson I've learned: don't get to know your heroes too well." All for what? So many well-intentioned people, once they become aware of this ugly scene, have tried to make it better. They want to right the wrongs and move on from the dirty laundry of the past to create a more effective future where everyone really can work together on behalf of these endangered whales. But for some reason, nearly all these efforts have failed. People have been hired as employees or become board members of regional organizations with the intent of mending fences with other groups, only to have their attempts derailed. People have started new organizations—research groups or naturalist networks, for example—that set out to have no affiliations, but those who don't take sides are viewed with mistrust and either forced into one camp or another or forced to disappear. The whale community has created a dysfunctional structure that insists we must often remain competitive, greedy, and territorial, just to do what we love. Of course, you can find similar politics in other arenas, but most people involved in the whale community agree that here it seems particularly magnified.

Not only is it frustrating, it's counterproductive. Cindy Hansen of Orca Network summarized it well. "If the whales are going to have a chance, we have to get past this," she said. "We're wasting our time, and the whales don't have time." After decades, many of these good-hearted people have become bitter and angry, transformed into someone they no longer recognize, so much so that some have left the whale scene entirely—they've moved away, or changed careers, or can no longer bear to be out there with the whales. Throughout the region are people once dedicated to the whales that have removed themselves from the whale community altogether. I have been there, and anyone who has been part of this scene long enough eventually gets there too—to a place where we are so exasperated we're on the brink of giving up what we love so much—the Southern Resident killer whales—because our fellow humans have made it such an intolerable place.

The term "orcatics" refers to local whale politics, but it really describes the human politics that surround the whales. Meanwhile the true whale politics—the politics of the Southern Resident killer whales themselves—starkly stand before us, telling a very different story. Perhaps it's my naive human perspective missing a lot of what goes on among them, but the whales certainly seem to resolve conflicts without violence, maintain camaraderie despite disagreements, and share grief without losing sight of joy. For forty years we've had an alternative model of how to behave right in front of us, from the very whales around which we've created our human web of deceit, betrayal, mistrust, and disappointment. The Southern Residents have survived loss, declining prey, changing group dynamics, and living with their immediate family twenty-four hours a day, 365 days a year, and they do it with grace, intelligence, love, and perseverance.

Consider L87 Onyx, the whale who has seemingly lost everything. He lost his mother, and then for some reason he left the L12 subgroup, with whom he had spent every day growing up, and started traveling with K-Pod. This was the first time a whale was documented changing pods. One by one, the females he grew attached to as surrogate mothers passed away as well, and he switched again from K-Pod to J-Pod. Onyx has a found a way to survive when many adult males don't after losing their mothers. He continually seeks out new companions, navigates the politics of changing groups, and still finds reason to play, regularly breaching, tail slapping, and swimming upside down at the surface in what looks like pure joy.

Consider J31 Tsuchi, a young adult female who has always loved calves. She often leaves her two brothers to hang out with families that have young ones, and I anxiously waited for her to have her own first born. When I saw the video in early 2016 of Tsuchi with her first observed calf, it brought tears to my eyes—she had given birth to a stillborn or a calf who didn't survive long. The footage showed her carrying the body of her little one on her rostrum. We've seen this from other grieving mothers, and we know sometimes they take their lost calves with them for many hours or days. My heart broke for Tsuchi, that she had such a sad experience with her own calf after finding such joy in the babies of others. Despite her loss, however, she spent the summer in the proximity of other young mothers, babysitting and playing with their calves.

I'm sure the Southern Resident killer whales have conflicts, as any complex society must. But I can't help but believe these whales have a much more harmonious community than humans do. What would they think if they had any idea of the insanity we have created around them? Since their population has survived all of the terrible things humans have put them through, I continue to look to the orcas for hope. If they can do it, maybe one day, so can we.

LEARN MORE

For more on the regional whale organizations mentioned in this chapter, visit their websites: Center for Whale Research (www.whaleresearch.com), Killer Whale Tales (www.killerwhaletales.org), Orca Behavior Institute (www.orcabehaviorinstitute.org), Orca Network (www.orcanetwork.org), Pacific Whale Watch Association (www.pacificwhalewatchassociation.com), The Whale Museum (www.whalemuseum.org), and Whale Scout (www.whalescout.org).

CHAPTER 8

The Most Watched Whales in the World

THE FIRST CONCENTRATED RESEARCH EFFORTS on the Southern Residents began in 1976, and it was just one year later that whale-watching started in the region. Only a few boats operated in the early years, but by 1988 there were thirteen vessels engaging in regional whale-watching, taking out an estimated fifteen thousand passengers a season. Throughout the 1990s, interest in whale-watching skyrocketed, and since then there have been more than thirty companies operating between seventy and ninety active whale-watch vessels each year in the Salish Sea, taking out hundreds of thousands of passengers annually.

This huge industry has led to the Southern Residents being considered the most-watched whales in the world—a direct result of how urban their environment is. Whale-watch trips depart not only from the San Juan Islands and Victoria, but also from Anacortes, Port Townsend, and Seattle in the United States and from Sidney, Cowichan Bay, Nanaimo, Sooke, and Vancouver in Canada. The boats are as varied as the companies that operate them, ranging from small six passenger boats to enormous vessels that take out up to more than two hundred passengers in a single excursion. To the whale-watcher, the fleet of vessels in the Salish Sea is as familiar as the whales themselves: boat names like *Supercat*, *Ocean Magic*, *Peregrine*, and *Explorathor* are as well-known as orca names Granny, Oreo, Blackberry, and Onyx. Shore-based whale-watching has drastically increased in popularity as well. While Lime Kiln Point State Park has long been known as one such site, there is now an extensive network of shore-based viewing locations throughout the Salish Sea. The Whale Trail is a series of onshore sites from which the public can view whales along the entire West Coast. Orca Network offers an extensive map specifically for Puget Sound.

As someone who has spent many years observing Southern Residents both on the water and from shore, I know firsthand that both experiences can be amazing, although each has its pros and cons. Boats can take you to the whales wherever they are, but they have to stay certain distances away. On land, you may have to wait much longer for the whales to come to you, but in the deep waterways of the Salish Sea you may be rewarded with an incredibly close encounter. When it comes to viewing wild animals, this is an unparalleled experience. Some people strongly advocate for one type of whale-watching over another, but the truth is, any experience with orcas is a memorable one, no matter where you're watching.

HOW MANY IS TOO MANY?

It was two in the afternoon on a Friday in July and I had just gotten off work. I stopped in town for a sandwich, where I ran into Nancy, a whale-watch captain. She didn't break

THE MOST WATCHED WHALES IN THE WORLD 215

her stride and said, "Sorry, I'm just between trips!" She had back-to-back whale-watching trips and a short window in between to grab a bite. Before she disappeared, she turned and said, "Js just got to Kellett when they flipped—it's a west side shuffle kind of day!" This sentence might not mean much to the uninitiated, but it turned my whole afternoon around. J-Pod, the whales who had been heard on the Haro Strait hydrophones earlier in the day, had reached the north end of San Juan Island and then turned around. This put the pod back within range for me to see them from shore on the west side of San Juan Island.

Half an hour later, I had eaten my sandwich and negotiated my way back to my car and through the tourists crisscrossing the streets in the main part of town. As I crested the hill in the Hannah Heights neighborhood, I got my first view of the water on the island's west side, and looked left along the southern part of the shoreline to see if I could spot any whale-watching boats. A cluster of boats right off of False Bay all faced the same direction. That split-second look told me that the orcas were in a tight group heading north, right toward where I planned to be. I was the only car in the north pullout at Land Bank's Westside Preserve. That would change as soon as the whales came into view, but for a moment I was the only one anticipating them. I grabbed my backpack (which doubles as a camera bag), my handheld VHF radio, and my hat out of the trunk before walking down the sloping hill with yellowing grass toward one of my favorite perches on the rocks about twenty feet above the water.

It didn't take long for the boats to come into view as they made their way around the point to the south. I'm a strong supporter of responsible ecotourism, but the sight didn't please me. There were more than twenty commercial whale-watching boats from both the United States and Canada clustered around the group of whales, each trying to edge in right to the agreed-upon quarter-mile buffer they give the whales along the shoreline here. The fact that so many people want to see orcas in the wild is a good thing, especially considering the education that naturalists provide, but it looked a lot like celebrities being mobbed by the paparazzi. I saw the boats a good fifteen minutes before I saw the first sign of any whale. By the time the first whale approached my spot on the shoreline, other people had figured out that an encounter was imminent. I turned, surprised to see that thirty or forty people crowded the shoreline above me. Many eagerly pointed or took pictures as the first whale passed. The orca surfaced in front of me about two hundred yards offshore. The half moon–shaped notch in the dorsal fin told me it was J2 Granny, often in the lead when J-Pod travels through Haro Strait.

There were thirty Southern Residents present—all twenty-five members of J-Pod and five L-Pod whales that traveled with them that summer. The animals were all within about half a mile of each other, though somewhat spread out. Some were offshore, too far away to identify in the harsh afternoon sunlight, but after Granny another group swam in closer to the kelp. I wasn't surprised to see J8 Spieden, J19 Shachi, and J41 Eclipse, three whales who often traveled close together and never roamed too far from Granny. Like Granny, the trio moved north at a good clip, but they had a little bit more of a playful attitude. Shachi tail slapped and her daughter Eclipse rolled onto her side at the surface, swimming sideways with her pec fin in the air for a moment.

Despite many regulations and guidelines, a mix of commercial whale-watch, private, research, fishing, and enforcement vessels surround the Southern Residents on an August afternoon as they move north up the west side of San Juan Island.

Surrounded by all the people watching the whales, I listened to their observations. "Are these orcas?" Somebody above me on the rocks asked. "Yes, but I didn't think this was the right time of year to see them here." A third person responded: "I think they're migrating north right now." Another asked: "I wonder if Ruffles is here?" And then: "Look, that group of whales is turning around. I bet they're turning around because there are so many boats here." The onlookers didn't have their facts all straight, but in the moment that didn't matter. I looked out to the boats to survey the scene. I knew from experience it always looks worse from shore than from the water (distances over water are incredibly hard to judge), and that the whales don't obviously change their behavior even with many boats present, but on this day it did look bad. I counted thirty-seven commercial and private motorized boats around the whales—more than one boat per whale— and on top of that more than a dozen kayakers.

One group of whales had indeed turned around, and as the whales spread out, the boats did too and the marine radio at my side flickered to life. The jargon of the whale-watch captains takes a bit of deciphering to understand. "They've stalled out at the Light," reported one captain, meaning one group of whales near the lighthouse had stopped traveling north. "I've got the leaders at 14, still northbound," responded another, using the numeric code the whale-watch companies have agreed upon to indicate that location. "There's a nursery group back here off the point just milling. The big guy offshore is doing some fishing." The captains' banter continued to crackle through my radio. "Good, it's shaping up to be one of those nights. I've got a turn-and-burn at six o'clock. How're the seas at the waterfront?" Another captain responded: "Threes, the occasional four, nothing bad. It's flat between here and there." In addition to discussions of whale locations, direction of travel, sea conditions, and planning for evening trips, the captains also discuss specific orcas. "Has anyone seen Blackberry?" "We've got him up here with Onyx, about 350 off my starboard bow."

Over the next hour, the whale-watch boats peeled off one by one and headed back to their home ports. The people clustered around me on the shoreline headed back to their cars. There were still whales in front of us, but the boats had reached the end of their trip times, and the shore-based watchers, having seen the whales, were ready to move on. Experience told me not to leave—for whatever reason, it seems nine times out of ten the best moment of a whale encounter comes when almost everyone else has gotten tired of watching and waiting and leaves. Before long, it was just me and the whales. The vision of the fleet clustered around J-Pod still in my mind, the scene had totally changed. Some anti-whale-watching people paint pictures of the whales constantly mobbed by their adoring public in boats, unable to feed or travel or play, assaulted by loud engines that accompany them on the water. Yet for about half the year, there are rarely whale-watching boats with the whales. Even now, during the peak season in the Salish Sea, the crowd of boats was condensed into just a few afternoon hours. From my onshore perch, I noted that some of the whales surfacing in the middle of Haro Strait didn't have a boat within a mile of them. A mom and calf surfaced with nothing but the Olympic Mountains behind them. Above me, the only other observers were three women sitting on the guard rail by the road, binoculars in hand.

A loud *kawoof!* broke me out of my reverie. I was surprised to see a whale in Deadman Bay swimming back south toward me. I turned to see if the women were going to come down closer to get a better view, but they were gone. I was the only one who was going to see this. It was J34 Doublestuf, one of the first whales that had passed heading north. I figured he would be heading back this way at some point, because his mom and younger brother never went all the way north. If the pod split, I knew Doublestuf wouldn't go north without them. He surfaced again, right in the sun track across the water, and I could only look in his direction because I was wearing my polarized sunglasses. Somehow, completely silhouetted, it was easier for me to appreciate the size of his dorsal fin. He was only fourteen but well on his way to having the tall dorsal fin of a fully adult male. As he glided by underwater, I made out his gray saddle patch through the green water and followed his movements from above the surface.

More than three hours had passed since I had scrambled down the rock to this viewing spot, and I was thinking about heading home to make dinner. Most of the whales had headed back south, and I figured the small group way offshore to the north of me probably accounted for the rest of them. I put my camera and binoculars away, but took one last look close to the shoreline. My timing was perfect: I saw a large splash. There were a couple whales porpoising back south. One of them was Granny. She'd come back to rejoin the rest of her pod.

Two whales broke off from Granny and slowed down, heading toward the cliff where I sat. It was L72 Racer and L105 Fluke. Racer swam under Fluke and pushed him partway out of the water. He was upside down, looking up at the sky. What did he think of the view? It was a tender moment between mother and son. After the pair passed by, I walked back up to my car where there was one other vehicle in the pullout. An older couple stood still, having spotted Fluke and Racer swimming together. "What a special thing," the woman said to me, with a smile. I tried to think of something to say, but nothing came. She had described the moment perfectly, so I just smiled back.

THE DEBATE OVER THE IMPACT OF WHALE-WATCHING

After resigning as the first executive director of The Whale Museum in 1980, Mark Anderson continued to make his home on San Juan Island and follow the plight of the orcas. In the late 1990s, the Southern Residents experienced a 20 percent population crash, from a high of ninety-seven in 1996 to just seventy-six by 2001. Anderson was concerned enough to found the Orca Relief Citizens' Alliance. Dedicated to understanding and reversing the reasons for Southern Resident population decline, Orca Relief's primary emphasis has become the negative impacts of whale-watching vessels via noise and disturbance. Explaining why his organization has taken this stance, Anderson doesn't deny that the main reason the whales are dying is because of salmon declines. However, he points out that any recovery efforts to increase prey or decrease toxins are efforts that will take decades to show any real effect. By contrast, vessel disturbances are something that can be mitigated immediately. "Also, there are many other groups working on toxins and fish," he said. "We have chosen to focus on noise and disturbance because so little has been done in that area, and we are a small group of modest means."

But what impacts exactly are whale-watching boats having on the whales? Orca Relief argues that engine noise masks the echolocation whales need to find food, that the presence of vessels around the whales results in behavioral disturbances that cause the orcas to spend more energy, and that both of these result in nutritional and psychological stress on the animals. Orca Relief's website is an excellent resource for finding the scientific studies that have looked at the impacts of vessels on killer whale behavior, but no strong links have been made between vessel disturbance and whale survival. The organization believes one solution is the creation of a no-go zone for boats on the west side of

San Juan Island, where the Southern Residents have historically spent a lot of time in the summer. Their initial proposal to NOAA to create an exclusion zone for all boats
was denied, but Orca Relief is petitioning for the creation
of a protection zone that applies exclusively to commercial
whale-watch boats. The organization argues that whale-
watch boats are the most problematic because they spend
the most time near the whales. Opponents point out that
shipping traffic, while further away from the whales, is actually a much greater source of noise disturbance and would
not be addressed under this solution.

*Whales do what they want, sometimes surfacing amazingly close to vessels of
all types. Some people claim even quiet, slow-moving kayaks disturb the whales'
behavior, but others argue that they are no more disturbing than a log floating
in the water.*

Shore-based whale-watchers can be among the other critics of whale-watching boats. Distances can be deceptive
from shore when looking over water. Often shore-based

viewers are outraged that boats are "right on top of the whales" when in fact, as required, they are well over two hundred yards away from the animals. On the water most captains do their best to follow the whale-watch guidelines and regulations, but it's true that at times the scene can get ugly. If there's only one group of whales around during an afternoon in the peak season, most of the fleet might be with the same whales at once. From shore, seeing boats vastly outnumber whales doesn't look good; even if everyone follows the rules, it just doesn't feel right. Management of this is difficult, when there are so many companies with different home ports, type of boats, and trip times.

Members of the Pacific Whale Watch Association must balance giving their customers a good experience with protecting and respecting the whales. Most company owners take the educational role they play of introducing the public to the whales very seriously. One example is Brett Soberg, co-owner of Eagle Wing Whale and Wildlife Tours in Victoria. A Victoria native, Soberg got hooked on the whales in the mid-1990s after getting invited on a free whale-watching trip while working at a hotel. After several years in the industry he cofounded Eagle Wing with the goal of building a sustainably responsible corporation. They've been successful, as Eagle Wing has won numerous awards for sustainable tourism and environmental leadership. Their business has grown into a carbon-neutral whale-watch company that uses a "wildlife fee" added on to their ticket prices to donate tens of thousands of dollars to environmental nonprofits each year. They're just one of many companies that do so, demonstrating the additional ways whale-watch companies can benefit the whales beyond education, raising awareness, and inspiring action.

THE EFFECTIVENESS OF NATURALIZING

For six years I worked as a naturalist with a whale-watching company out of Friday Harbor. I wanted to share my love of the Southern Residents with others in hopes that they would be inspired to help protect this endangered population. Most people agree that the regional commercial whale-watch industry has a huge opportunity to provide this sort of education to the thousands of tourists from all over the world that come out to see the orcas each year. Naturalists may play many roles, from safety instructor to entertainer to server to babysitter. A parent may expect you to police their child and keep them from climbing up on the boat rails. Someone else may interrupt your discussion of killer whale family trees to ask for hot chocolate. Other people, being on vacation, are quite adamant that they're only there to have fun and are not interested in learning. It's somewhat of an art to balance all of these things and still fulfill the goal of education. Results vary.

You can't hold it against somebody that they don't have knowledge about whales before a whale-watching trip, but I hope that they will at least leave a whale encounter with more information than they started with. It's easy to feel deflated after your best efforts of providing engaging education are met at the end of the trip with this question: "So what kind of whales did we see?" Naturalists enjoy comparing the outrageous questions we have been asked. Some classics include: What time do the whales come by? Can you make the whales come closer? Can you make them jump? Why are they following the boats? Do the whales come out when it rains? Do whales give birth in the water?

Thankfully not all, or even many, interactions are like this. Naturalists' educational efforts can also be extremely rewarding. Perceptive and engaged passengers often ask intelligent questions that lead to fulfilling discussions. I once had a passenger on board who was a chemist. She was very interested in the research on the detrimental effects of toxins on marine mammals, and in return I learned more about the chemical nature of some common pollutants. Another time a child asked me what "endangered" meant and it led to a discussion about endangered species recovery that included everyone on the boat. In these instances, naturalists get a real opportunity to converse about what is and isn't known about Southern Residents, about the research that has led to what we think we do know, and about what can be done about the conservation issues facing these whales. It's these interactions—where both the passenger and the naturalist are enlightened by new perspectives—that truly make the naturalists' profession a rewarding one.

Pete and Nancy Hardy are a husband-and-wife pair of captains who work as relief drivers for whale-watch operators in the San Juan Islands. In the summer they live aboard their fishing boat moored just outside Friday Harbor. In addition to being captains, they are also both excellent naturalists and feel strongly that the focus of the trips should be on more than just orcas, despite the fact that they are the main species advertised and the primary goal for most of the passengers. When whale-watching first started in the region, the focus was on all types of wildlife but that shifted to becoming very orca-centric. Recent workshops among naturalists and captains in the whale-watch industry have discussed returning to more wildlife-focused advertising, but as long as one company "guarantees" whale sightings (handing out vouchers for a free return trip if you don't see them) and continues to feature breaching killer whales on brochures and social media, it's difficult for the other companies not to follow suit.

"People come here to have an enjoyable vacation," says Pete, "and we have the potential through education and entertainment to show them how special *everything* is out there, not just the whales." In fact, they've coined a term—"interpretainment"—for what it is they think a naturalist really has to do. Not just interpret, and not just entertain, but to be effective, a combination of both. Pete believes that by and large naturalists are doing a good job and that the whale-watch industry is playing a big role in the conservation of the Southern Residents. Nancy, however, isn't quite as sure. "We only spend a few minutes of every trip, really, talking about conservation," she points out. "Most naturalists present the information well, but I'm not sure if people really get it." Pete says educators should take it as a challenge to make each passenger see things a little bit differently than they otherwise would. It might be hard to get through to a person who just wants to check seeing a killer whale off their bucket list, but naturalists should still try to open their eyes to at least one new thing. "I don't think we really see the impact seeing whales has on people," Pete muses. "The influences are probably small and immeasurable, and may occur well after the trip is over, but I believe that they're there."

All naturalists have seen passengers get off a whale-watching boat profoundly moved by what they've witnessed. One naturalist said he continues working in the industry for those moments when the whales bring people to tears. Soberg of Eagle Wing said everyone is always riding a high after a whale-watch experience, and the trick is to tap into that high. He summed it up well: "We want to figure out how to maintain a connection to our clients long term, over social media or e-mail. This will keep them excited about the whales, which will both encourage them to come back and go out whale-watching again and to take action on behalf of the whales. Right now, I know whale-watching is having a positive impact, but there aren't metrics in place to measure how much of an impact. We need to figure out how to measure it, not only to show the good our businesses are doing, but because if it's not a big enough impact, we need to figure out what we can change to do better."

SHORE-BASED WHALE-WATCHING AND THE LOST ART OF PATIENCE

In the Salish Sea you don't have to take a boat to go whale-watching, but to see a whale from shore can take a little (or a lot) of patience. Today's society is not accustomed to waiting. If we wonder about something, we Google it, answers just a few keystrokes away. Virtually all our media—movies, music, TV shows, even games, books, and magazines—are available "on demand" with no need to leave our homes to acquire new entertainment. We can reach people far away without delay, almost no matter what they are doing, via an email, text message, or video chat. Even some of our greatest natural wonders are predictable scheduled events, leaving nothing to chance. Old Faithful will next erupt at 10:00 a.m. The meteor shower will peak at 11:30 tonight. The app on your phone will tell you to the minute when the sun sets, so you can plan accordingly, no waiting required. Marine theme parks do the same for wildlife: If you want to see an orca jump, come to our next showing at 2:10.

Many people carry what some naturalists call "the SeaWorld mentality" to the shores of the Salish Sea, where wild orcas frequently swim by during the summer months but on their own schedule, not ours. Some people come with the mind-set that the whales, wild or captive, exist to perform for them and therefore should appear on a predictable schedule for their viewing pleasure. When they do get to see whales, they're often not impressed with "just breathing" and want to see behaviors like breaching. Of course, not everyone acts this way. For some, the magnetism of wild orcas is enough that it breaks them out of their trance of expecting everything right here, right now. One special place to watch this play out is at Lime Kiln Point State Park. During summer, orcas can pass through on nearly a daily basis, sometimes several times a day. For the patient observer, Lime Kiln is the place to be. Sometimes it's only a small pause, such as when a group of restless children who just happened to show up at the right time suddenly sit still, in silence, waiting for the next surfacing. There is a small wonder even in this—knowing the whale is out there, underwater and out of sight, but not knowing where it will come up next, and waiting, breath held, to find out.

Then there are those people who will wait hours or even days for a whale sighting. They'll sit on the rocks at the Lime Kiln Lighthouse with no screens or internet connections or cell phone coverage and just wait, not knowing but hoping that the whales will show up. They aren't impatient as time ticks by; they have found a different sort of calm— sitting in the sun, feeling the wind, watching the tide come in and go out, and reawakened to the fact that waiting is not really waiting at all but observing. Lime Kiln is nicknamed Whale Watch Park, but it's perhaps more appropriately known among locals as Whale Waiting Park. Waiting is not a bad thing; it is in fact something humans were born to do and it can put us in touch with our surroundings.

I talked to Bob Otis about this, the researcher who has conducted shore-based whale research at Lime Kiln Lighthouse since 1990. Every day between May 20 and August 10 he's at the lighthouse from 9 a.m. to 5 p.m. to collect data if the whales pass through his study area. He is the ultimate whale waiter and has seen more than anyone how people behave in the park while looking for whales. "We've lost the observer mentality," Bob said. "Many people aren't content to just sit and wait and watch. We even see that here in the Park. Before the pagers and text messages, we used to have someone outside looking for whales all the time. They would see all other kinds of amazing stuff. They would call everyone outside to see a family of otters, or a gull trying to eat a starfish. Think of how much we miss even today while we wait for the whales."

"But a lot of people *do* come to Lime Kiln and simply wait and watch for whales," I pointed out. I think of the interesting cast of characters I have met over the years, some of the most dedicated whale waiters, and people Bob knows too. "Someone told me they find Lime Kiln so exceptional because it's a place where people go to wait for something to happen. How often are people content to wait for things in this day and age?"

Bob nodded. He knows all the whale waiters I do, plus many more. "The really good things in life are fleeting," he said, "and the whales are one of those things worth waiting for. One reason I have hope for the future of these whales is the number of kids who want to wait to see them. The parents want to go and see the next thing on the list, but the kids want to sit here and wait for the whales. There *are* people out there who get it."

There's an even higher level of patience when it comes to waiting for wild whales—those who wait years just to have the opportunity to stand on the rocks at Lime Kiln and see one. I'm amazed at the number of people who see my photos and say something like, "One day, that will be me. Seeing a wild orca, just once, is on my bucket list. I hope to plan a trip there in about two years." When one of those people, a woman, comes after years of waiting, she finally finds herself at the lighthouse. It's not quite like she pictured: it's raining, and fog clouds more than half the strait. But there they are, a dozen wild orcas. Not close to shore, not breaching, but it doesn't matter. This is an experience you can't get anywhere else. Happy tears slip down her cheeks, indistinguishable from the mist that dots her glasses. J2 Granny surfaces, her breath loud on this quiet morning. Surrounding Granny are her family—grandchildren and great-grandchildren. They swim south, going to a place only they know, for reasons we can only speculate. The vast ocean is their realm, and the whale-watcher is granted this one brief look into their lives. This was a moment worth waiting for, and it played out at Lime Kiln.

Just like the on-the-water community among members of the Pacific Whale Watch Association, there's a community of shore-based whale-watchers that centers around Lime Kiln. People from all walks of life come to spend hours, days, weeks, or even months waiting for encounters with the Southern Residents. What follows are portraits of a few of those people.

LIME KILN WHALE-WATCHERS

Sjors can spend twelve hours straight looking through binoculars. He's perfectly happy sitting on the rocks, hour after hour, scanning the water for any sign of marine life. You see some amazing things if you look at the ocean that long. He's found orcas that a fleet of whale-watching vessels has zoomed past, male elephant seals with only their gigantic snouts breaking the surface, and a lone humpback whale miles from shore, during years when this was still a rare sight. Even while he's sitting and telling stories about how many times he's been stung by a scorpion (nine) or the time he was dropped off on a remote island to do a gray whale census, he pauses every few minutes to search the waves for anything new: a spout, a fin, or a back.

Many people hope for a moment like this at Lime Kiln: a wild killer whale (in this case, J34 DoubleStuf) swims by just yards off the rocks.

Sjors wears a white baseball cap with a custom extension to further block the sun pulled down over his squinting eyes. During the year he lives at the bottom of the Grand Canyon at Phantom Ranch. He's mostly a volunteer, there to warn confident hikers that no, they really should wait a few hours until the temperature drops below 120 degrees before hiking back to the rim. When they don't listen, or get lost on the way, it's Sjors's job to help, whether that means bringing someone water or going on a helicopter search-and-rescue mission. I've always wondered why someone who loves the ocean and whales so much lives at the bottom of the Grand Canyon. Every once and a while I get an email from him that gives me a glimpse into why that might be. He's seen the northern lights from the canyon floor, felt snowflakes while standing on the banks of the Colorado River, and watched mating displays of wild California condors. Still, as much as Sjors enjoys the exhilaration of his life at Phantom Ranch, he climbs out of the canyon once a year to return to the orcas.

Another devoted whale-watcher is Shane, who lives with his dad on Orcas Island, a short ferry ride from San Juan. When his father has some errands to run or has some time to accompany Shane on the ferry, he'll drop Shane off at Lime Kiln for the day. Shane brings the same two items with him on every visit: an old pair of binoculars and a harmonica. He is truly happy when he gets to talk about one thing: orcas. Some people don't have the patience to talk with Shane because of his Down Syndrome, but if you join him for a while, you'll realize how much he understands the whales. He knows the different behaviors and some of the individual whales. Even if no one is around to talk to Shane, he'll keep himself busy. He plays the song from *Free Willy* on his harmonica to help call the whales and glances at the horizon through his big binoculars to see if he can spot any "big splashes"—his telltale sign of the whales coming.

I sent Shane a movie I put together of the killer whales, and he has never forgotten it. His dad tells me it's the only movie besides *Free Willy* that Shane will watch again and again. When I see him after a year's absence, Shane shakes my hand, pats me on the shoulder, and jabbers excitedly about the last time he saw Ruffles, his favorite whale. When his speech becomes hard to understand, he'll revert to sign language. He has taught me some very important signs:

beautiful, splash, and the unique sign he's made up for Ruffles. As excited as he gets talking about the whales, however, when they are present, Shane is perfectly happy to sit in silence, his eyes wide.

Bobby, another man who makes an annual trek to Lime Kiln, initially comes across as a stereotypical redneck. His speech is a gravelly drawl, he wears a dirty leather jacket with a trucker hat to cover his sunburned face, and he spends his evenings downing pints at Herb's Tavern in Friday Harbor. During the school year, however, Bobby teaches troubled kids and helps them prepare for reentering the public school system. And during the summer, he's a whale-watcher. Ever since he first saw the Southern Residents, he incorporates images and video clips of them to inspire his students about how much beauty is out in the world beyond the kids' difficult lives in Lubbock, Texas. Bobby makes the drive from Lubbock to the Washington State Ferry line in thirty-six hours, with only two rest area stops on the way for a quick nap and a cup of coffee. He spends entire days waiting for the whales, just like the rest of us. When the orcas aren't around, he'll spend time reflecting in his journal or chatting with other visitors. When the whales are in view, Bobby will climb as far down on the rocks as he can go, right where I like to be, with his disposable camera in hand as I snap shots with my Nikon DSLR.

He wants to bring his own kids to Lime Kiln some day to share the experience. I think he hopes I will befriend his oldest son, because every year Bobby encourages us to start an email correspondence. He tells me how smart his son is and how he's into biology, just like me. The one time I did email

Bobby's son, he complained about how I was the reason his dad refused to buy a DVD player. When Bobby's VCR broke, he insisted on buying another one so he could play his VHS copy of my old whale video as a pick-me-up. Although his son still hasn't seen the whales (except on his dad's new VCR), Bobby keeps coming back to see the real thing.

When passionate orca enthusiast Jeanne first arrived on San Juan Island, I don't know if she knew what she was getting into. Needing a new direction in life as she approached sixty, Jeanne sold her house in the South, rented an apartment in Everett, Washington, and planned to spend the weekends on San Juan Island looking for killer whales. A few days into her first visit to Lime Kiln, she decided to volunteer. A few weeks later, she was living in her tent at the state park four nights a week and only briefly returning to Everett. A month or two after that, Jeanne gave up her Everett apartment entirely. She later boasted about not having left San Juan Island for any reason for nearly two years straight (she finally had to go off island to get new tires). It wasn't long before Jeanne was interested in identifying the whales. She upgraded her camera, bought ID guides, then upgraded her camera again. She continued working as a docent in the park, where she spent all her time anyway, and landed the job of Orca Adoption Program coordinator at The Whale Museum, a perfect fit to accompany her increasing knowledge of the whales and love for telling stories about them.

Like many shore-based whale-watchers, Jeanne worries about missing one of those extra-special encounters on the west side. Making plans with her is footnoted by the statement "whale dependent," meaning she'll be there unless there are whales to be seen. She now puts in her work at The Whale Museum at night—you can't see the whales under the cover of darkness, but you can listen in online over the streaming hydrophones. She leaves the hydrophones on as she sleeps, trusting herself to wake up should any eerie killer whale vocalizations break the silence of night.

Bundled up against the wind, Jeanne heads down to "her rock" at Lime Kiln with all her equipment. Sitting with her is an experience in itself. When the whales pass by, she takes hundreds if not thousands of photos (deleting none of them later to avoid losing part of the story). If the conditions aren't right for photos (for example, very early in the morning or in the fog), she shoots video. One ear is tuned in the hydrophones broadcasting live, and often the other is busy talking on the phone with her growing network of whale contacts, including researchers and boat captains. She has become a great source of information for the sightings networks. Amid all this, Jeanne still manages to call out the whales by name as they pass by; she can identify them on sight as well as anyone and better than most. In the summer she keeps a daily blog of her whale encounters, sharing pictures and stories of every Southern Resident in the community. In the winter she pours over her photos, memorizing every new nick and scratch on every whale. She makes products out of her images—note cards, buttons, and family ID booklets. And, as always, even in the off-season, Jeanne is ready to jump at a moment's notice to see any orcas that pass through the area.

In 2009 the Samish tribe held a naming ceremony for J14 Samish's new calf, J45. The Whale Museum selected Jeanne to participate in the ceremony on their behalf, and part of her role as an honored participant was to share the story of this newest member of the Southern Resident community, whom they named Se-Yi-Chn, meaning "younger sibling." They couldn't have picked a better person than Jeanne, who has dedicated her life to knowing the whales and telling their stories, to honor with this privilege.

Seeing the whales changed the course of Julie's life. A divorced mother of four who had always loved nature, she was inspired by the orcas to return to school and pursue a degree in biology. Despite her studies she always makes time for an annual summer pilgrimage from central California to San Juan County Park, where she camps for two weeks and watches whales. She brings some friends with her, but she makes others at the campground. Her outgoing and friendly personality transform the separate groups of campers into an interactive camping community. Julie's campsite is always the center of activity, bringing together everyone from Germans touring North America by bike to LA city girls in the name of peach margaritas and campfire talk about whales.

While Julie enjoys watching the whales from both boat and shore, her favorite encounters come by kayak. Never selfish, she always brings two kayaks with her from California so she can share the experience with the new acquaintances she knows she'll make. Unlike many whale-watchers, Julie never takes photos, preferring instead to sit in the kelp, bob in the waves, and just take in the whole experience. When friends and family from back home clamor for pictures of what keeps drawing her back to the island year after year, she doesn't give in and bring her camera the next summer. Instead, she extends an invitation for them to join her and experience it for themselves.

After completing her undergraduate degree, Julie decided to get her PhD. The assumption might have been that she would want to study killer whales, the creatures that motivated her to learn more about animal behavior in the first place. Instead, she chose to study tuco-tucos, a rodent found in Argentina that inspired another annual journey to South America every winter. Julie wanted the magic of her orca encounters to remain untouched by the sometimes-esoteric world of academia. She intuitively knew that equations, experiments, and scientific journal articles would never fully capture what the whales meant to her, so rather than tarnish the experiences in the name of research, Julie left them to be mysterious and inspirational, meant to be fully taken in doing nothing but sitting in a kayak on the edge of their realm.

While I don't know for sure, I suspect Joe isn't your typical naval officer. He spends most of his free time on deck, scanning the ocean of wherever in the world he happens to be for marine mammals. He sees a lot of whales and dolphins that way, but none of them can ever quite match his passion for orcas. As he gained seniority in the Navy, Joe's request to be stationed at the Whidbey Naval Air Station was granted. Whidbey was his top choice because of its vicinity to the Southern Residents. Although most of his whale-watching has been done by boat, Joe prefers to see these whales from shore at Lime Kiln Point State Park. Scanning Orca Network reports for recent sightings, he requests leave to spend a few days on San Juan Island.

On the rocks at Lime Kiln, the soft-spoken Joe blends in easily. He sits back from the bustle of activity with a ball cap pulled down low on his forehead. After he recognizes you as a kindred spirit when it comes to the whales, he chats amicably, but only until the whales show up. Then he retreats to his own spot to watch them. Oddly enough, despite years of waiting with Joe for the whales, I can't recall a single time I actually saw the whales *with* him. Perhaps he prefers to observe the whales in solitude.

Joe's wife and two sons don't quite share his passion for orcas—they'd get restless during an entire day at Lime Kiln. Because of this, Joe often comes alone, but other times he brings young men from the Navy. Usually they are officers, Joe explains, who are having some difficulties in their lives, and for one reason or another they are seeking meaning or direction. Not one of them, despite making their living on the ocean, had ever seen a wild whale. It was fascinating to see these men—sometimes heavily tattooed, sometimes chain smokers—sit at the park and wait for the whales all day. I never talked to them about it, but I wonder if seeing the whales helped them to find what they were looking for.

A PASS-BY FROM THE L12 SUBGROUP

The L12 subgroup is made up of the L12, L28, and L32 matrilines and also includes the older female L25 Ocean Sun. Ranging in size over the years from eight to seventeen whales, this portion of L-Pod sometimes spends more time in inland waters during the summer months than the rest of L-Pod, making them better known among naturalists. Throughout the 2000s they were particularly famous for the "west side shuffle," where they would spend day after day traveling slowly from Salmon Bank to about Lime Kiln before turning around and heading back south. For whatever reason, they rarely go north to the Fraser River, instead spending a lot more time in Haro Strait.

A classic L12 encounter took place one summer I sat on the rocks at the Lime Kiln Lighthouse with Bob Otis and his interns, Sara and Loryn. It was late afternoon on one of the few days each year where it's warm enough to sit right by the water in a T-shirt, the water glassy calm as it ran north on a strong flood tide, undisturbed by any whisper of a breeze. Usually when a summer pass-by occurs at Lime Kiln, there's a lot of activity going on. More tourists appear on the shoreline, somehow emerging out of the woods having sensed the whales' presence. I perch low on the rocks with my camera in hand, clicking away and monitoring my handheld VHF radio. Bob and his interns scribble notes on their clipboards and call back and forth to one another, noting whale IDs, the time and distance from shore when each whale passes in front of the lighthouse, and all surface behaviors that occur anywhere within a half-mile in any direction. A speaker, connected to the streaming hydrophones, plays the orcas' live vocalizations for all to hear. The whole thing usually lasts anywhere from about twenty minutes to an hour. This day, however, things were a little different.

The eleven members of the L12 subgroup had been "passing by" for the last two and a half hours. It started out like a normal pass-by, with the whales coming up from the south in two separate groups. Lots of people had gathered on the shoreline to watch. After about half an hour, the whales were just north of the lighthouse when, as they are known for doing, they turned around to face back south.

The animals spread out, apparently foraging, facing into the strong flood tide. Swimming slowly, they made little to no progress against the current, surfacing every time in almost exactly the same spot for the next two hours. Tourists, losing interest, began to drift away. Soon it was just the four of us researchers with the whales—a rarity for a summer afternoon. The speaker, emitting not so much as an echolocation click from the silent whales, had been turned off. Having already noted every whale present, Loryn and Bob had set their clipboards and pens down on the rocks next to where they sat. Sweat glistened on Bob's forehead, but he sat resolutely in the direct sunlight, unwilling to compromise his data collection by retreating to the shade where he wouldn't be able to see the whales should something else happen.

No one had said anything for about twenty minutes when Bob turned to me and said, "You know, I've never said it before about orcas, but these whales are getting boring."

"No, they're not," I said, defiantly. "You never know what they're going to do next." Bob just smiled. L22 Spirit surfaced, facing south. The sound of her breath was audible from a quarter-mile away. She sank back beneath the surface, the water creating a little V-shaped ripple where the tip of her dorsal fin had disappeared. A minute passed. Another minute passed. Spirit came up for another breath, still facing south, but appearing a little further to the north. "It's like they're swimming backward," Loryn said. It was true: they moved so slowly that they weren't making any progress in the direction they faced. They were actually drifting backward with the current.

Ten more minutes passed in silence. "They're spread out like they're foraging," I said, "but they don't seem very active." Bob had a theory. "They're probably just facing into the tide with their mouths open, hoping a salmon will swim in," he said, chuckling as he pictured this. It wasn't a bad speculation. What else were they doing? One curved dorsal fin broke the surface to the left. A moment later another orca, a male, surfaced offshore to the right. Then, again, nothing.

Sara had retreated about a hundred feet away to the flat rock where she was known for sprawling out to nap on days when there was little to do but wait patiently for the whales to show up. She lay on her stomach, facing the water, chin resting on her folded arms. One hand held her pen, and the clipboard lay on the rock in front of her. Her eyes were concealed behind sunglasses.

"I've never seen an intern collect data like that," Bob said, pointing at Sara. Loryn and I laughed. "Do you think she's awake?" Bob wondered. None of us were sure. Bob reached down and pulled his video camera out of the case at his side. Slowly he raised it to his eye and flipped it on, filming Sara. She made no movement if she saw us watching her.

Kawoof! I glanced back toward the water, hearing the loud sound of an orca breathing. A moment later, a second breath: it was L32 Olympia and her son, L87 Onyx, surfacing side by side. As Olympia dove, she gave a little nonchalant flick of her tail. We looked back to Sara, who had the data sheet for recording surface behaviors. She moved her hand forward, writing a quick tally mark on her data sheet. "She is awake!" Loryn exclaimed, loud enough for Sara to hear. She turned toward us, still not responding to the video camera trained on her.

"That's a pretty half-assed tail slap!" she commented. Bob laughed, but it was true. It was the closest thing I had ever seen to half a tail slap. Olympia and Onyx surfaced again, and again she half-assed a tail slap on her way down. She continued to do this on every surfacing for the next half hour. Later, when I saw Sara's data sheet, she had written in a separate behavioral category for Olympia. It read "HA tail slaps" and had its own row of tally marks lined up next to it.

Bob's data-collection protocol states that if a pass-by begins before 5:00, data collection has to continue until all the whales are out of the study area. Shortly before 5:00, Loryn had to leave to head to town to start her waitressing shift. She left her clipboard by my side to continue recording data if need be. She was supposed to record boat behavior, but by this time all the whale-watchers had departed to head back to port; she hadn't written anything down for almost half an hour. At 5:30 the L12s had made a little progress moving south. They had gone back and forth a lot, swimming south, then drifting back north in the current. "Do you think they're out of the study area?" Bob asked.

"I don't think so, not quite," I said.

"I mean, do you think they're out of the study area?" Bob repeated. "I never got to have lunch . . . "

"Oh . . ." I looked again, through binoculars. "It's pretty close. Sure." That was agreement enough, and we packed up our things and put the clipboards, radio, ID guides, and cameras back into the lighthouse. I flipped off the hydrophone that was still faithfully recording silence on the computer. I dreaded the task of going back through to listen to this recording in the off chance that some sporadic vocalizations had broken the hours of silence.

Bob locked the lighthouse and we walked up the hill back toward the parking lot. A *kawoof* made us all stop for a moment and turn around. Another whale had surfaced, clearly within the study area. Bob, Sara, and I glanced at each other, said nothing, then turned and continued walking away. It was the only time that summer I left the park when there were whales within sight, but Bob was right. The L12s were sometimes pretty boring.

LEARN MORE

For further in-depth stories about the profound ways orcas can impact the lives of whale-watchers, read *Orca: The Whale Called Killer* by Erich Hoyt, *Listening to Whales* by Alexandra Morton, and *Into Great Silence* by Eva Saulitis. To learn more about seeing whales in the Salish Sea, look up maps created by Orca Network and The Whale Trail for shore-based viewing sites, and visit the Pacific Whale Watch Association's website for a list of companies that engage in responsible whale-watching.

PART 3

ORCAS IN TROUBLE

CHAPTER 9

The Three Risk Factors

AT THE END OF THE live capture era, the Southern Resident killer whale population numbered about seventy animals. Over the next two decades, their numbers slowly increased to nearly one hundred animals by the mid-1990s. But in just five years, the whales experienced a 20 percent population decline, which led the Center for Biological Diversity to petition the National Oceanic and Atmospheric Administration (NOAA) to list the Southern Residents under the Endangered Species Act (ESA) in 2001. Also in 2001, the Committee on the Status of Endangered Wildlife in Canada listed the Southern Residents as endangered under the Species at Risk Act (SARA).

NOAA acknowledged that the whales were in trouble, but since killer whales are identified as a single global species that are doing well overall, the Southern Residents did not meet the criteria for an endangered listing. NOAA began the process to identify them as a depleted population under the Marine Mammal Protection Act and were also challenged in court to reconsider the endangered listing. NOAA's biological review team that assembled in response to the court order led to the designation of the

Southern Residents as a "distinct population segment" of the species *Orcinus orca*, which identified them as a separate and unique population of killer whales eligible to be listed under the ESA. The review process continued for several years, but in November 2005 the Southern Residents were at last listed as endangered in the United States. As a result of their endangered listing in Canada and the United States, both countries identified three main risk factors as hurdles to the recovery of resident killer whales: vessel effects, toxins, and lack of prey. None of these issues stands completely alone, but that doesn't mean they are equal threats.

VESSEL EFFECTS

Whether it's okay for whale-watch boats to outnumber whales is a moral question for us humans, but what has science shown in terms of impacts from boats on the whales? The topic has been researched over the past several decades, and some minor influences of whale-watch boats on resident killer whales have been revealed, mostly in the form of masking effects on their vocalizations, making it more difficult for them to communicate and echolocate. Behavioral responses, such as changing dive times and swimming paths, have been shown with close-range approaches by small vessels, but the effects differ by population, location, time of year, and age and sex of the whale, making any impact from boats harder to discern from overall behavioral variation. The energetic impacts of whale-watching on the whales are pretty small; as of yet, nothing links these vessel effects to population-level parameters such as survivability, reproductive success, or foraging success. Sometimes we forget these urban whales have long dealt with large numbers of boats; in fact, hundreds of fishing vessels plied Haro Strait before whale-watching even began. For better or worse, these whales are adept at handling the presence of many boats.

While whale-watch boats are the most visible vessels near the Southern Residents, these are not the boats having the greatest impacts. Other real and potentially major risks to the whales include the near-constant noise from commercial shipping traffic in their core summer habitat as well as from other activities such as drilling, construction, seismic testing, and sonar; the risk of vessel strikes; and the potential for oil spills. Commercial shipping worldwide has substantially increased the ambient noise levels in the oceans; one estimate shows that ambient noise in the northeast Pacific Ocean has gotten four times louder since the 1950s as a result of commercial shipping. More than ten thousand oil tankers, cargo ships, and tugboats with barges transit through the Salish Sea each year, with this number projected to grow in the coming years, particularly if proposed coal and oil terminals are built near the United States–Canada border. Much of the engine noise from these vessels is low-frequency and has been demonstrated to affect the ability of baleen whales to communicate in other parts of the world. However, in 2016 the team responsible for maintaining the hydrophones on the west side of San Juan Island published a study showing the shipping noise in Haro Strait extends to the mid- and high-frequencies used more by killer whales, potentially effecting their communication and echolocation. Despite this, the Southern Residents are known to sometimes swim directly at the freighters that share their habitat, either to pass right under them, or at times to surf their wakes, seemingly just for fun. The masking effects of ocean noise probably does have a real impact on the whales, but this at least raises the question of how much vessels are really directly bothering the whales.

A group of Southern Residents surfaces next to a freighter in Haro Strait. More than thirty of these loud ships traverse the area each day, and sometimes the whales seek them out to surf their massive wakes.

Both the Canadian and US militaries operate in the Salish Sea throughout the year, including testing midfrequency active sonar as well as abovewater and underwater ordnance and munitions. In the United States these federal activities have to get approval from the National Marine Fisheries Service (NMFS) because they may result in "take" on protected marine mammal species. NMFS regularly ends up involved in legal battles with such groups as Earthjustice and the Natural Resources Defense Council for failing to adequately protect marine mammals when issuing permits to conduct these exercises. Although there are procedures in place to help ensure these activities don't occur when there are marine mammals like killer whales in the area, there have been instances of testing occurring in the presence of the Southern Residents.

The most notable incident occurred in May 2003, when the USS *Shoup* operated sonar in Haro Strait while J-Pod was present. Coincidentally, J-Pod was right off the Center for Whale Research's Orca Survey house, being watched by Ken Balcomb, when the guided-missile destroyer activated a midfrequency sonar test. Balcomb captured astonishing video footage of J-Pod's reaction; they grouped up close to shore, spy-hopping and tail slapping in a bizarre behavior that he had never witnessed before. He also filmed a minke whale porpoising away from the source of the sound, another unusual behavior. Throughout the video, the intense sonar pings are audible, at times above the surface of the water. The short-term behavioral response proved to be the only measurable impact on the Southern Residents, but other marine mammals in the area weren't so lucky. In the coming days, more than ten harbor porpoises stranded in the area. Most were collected in various states of decay for necropsies, and various causes of death were determined. In recent years, the Pacific Whale Watch Association has been playing an increasing role in notifying the military when whales are entering an active testing zone, which will help prevent an incident like this from happening again.

The impacts of Navy testing exercises on marine mammals have been well documented in other cases around the globe. With ongoing testing and proposals for new operations throughout Southern Resident killer whale critical habitat, the risk of an incident involving military vessels and orcas remains real. Naval testing occurs not only in inland waters but also along the outer coast, where a bombing test zone ironically overlaps with the Olympic Coast National Marine Sanctuary. (Ken Balcomb once said, "A sanctuary from what?"). A single catastrophic event could easily spell extinction for these whales. The danger is indisputable. In February 2012, for instance, three-year-old female L112 Sooke washed up near Long Beach on Washington's outer coast. She showed significant internal injuries to the head and along her right side, and although the necropsy report identified blunt force trauma as the cause of death, officials failed to speculate on what might have caused these injuries. We are left to imagine the worst: ship strike or perhaps explosive?

L-Pod was in the Strait of Juan de Fuca during naval sonar activity a little over a week earlier, and from where Sooke washed up, it was theorized she could have died within the outer coast bombing zone. Balcomb, a veteran of the Navy himself, has been vocal about test explosives being the likely cause of death of this young whale. As a young female in L-Pod, where not many female calves have been born, Sooke was especially critical to this population. In late 2016, J34 DoubleStuf washed up near Vancouver with bruising and evidence of blunt force trauma, but this time, no official cause of death was given and, as of two years later, no final necropsy report was released, adding to the mystery of what killed these two whales. Military activities had been going on nearby in the days preceding his death.

There also exists the potential for killer whales to be hit by vessels or entangled in active or derelict fishing gear, although these incidents are rare. NOAA's recovery plan for Southern Residents, a strategy they were required to develop after the endangered listing and completed in 2009, notes that there have been three fatalities due to ship strikes among both Northern and Southern Resident killer whales between the 1960s and the present day, and one of these was L98 Luna, who had an affinity for vessels and was killed by a tug boat. Not all vessel strikes or entanglements prove fatal; there have been about half a dozen known cases of resident orcas being struck by vessels and receiving injuries from which they recovered. In August 2015, for example, A95 of the Northern Resident community was photographed with nasty-looking injuries deemed to be from a propeller, but the mostly superficial wounds healed quickly. In addition, earlier that same month, J39 Mako was photographed with a salmon flasher in or attached to his mouth, but five days later it was gone and he was observed in fine condition.

Another threat is the possibility of a large-scale oil spill. Thankfully the Southern Residents have not experienced one, but the risk is real, and will possibly increase with recent proposals for new coal and oil terminals in both Washington and British Columbia. In 2016 opposition by Northwest Tribes to what would have been the largest North American coal port at Cherry Point in Washington led to the US Army Corps of Engineers denying the permits after a long-fought battle. In 2018, a Canadian court ruling halted the controversial Kinder Morgan pipeline project, in large part because the concerns of tribes and impacts to the endangered Southern Residents were not adequately considered. The proposed expansion of the tar sands pipeline from Alberta to BC's coast led to a series of protests and court challenges, this last one being enough for Kinder Morgan to walk from the project. Despite this, Canada's Prime Minister Justin Trudeau promised construction would be completed and stated the Government of Canada would purchase the project if no other buyer could be found. If the expansion proceeds, tanker traffic in the Salish Sea would increase sevenfold.

Our knowledge of the effects of an oil spill on killer whales comes almost exclusively from the catastrophic 1989 *Exxon Valdez* spill in Prince William Sound, where two different groups of orcas were seen swimming through the oil. In the resident AB Pod, seven whales died shortly after the spill and six more over the following winter, totaling a third of the pod. In addition, two adult males showed dorsal fin collapse, which is unusual in wild whales and may be related to poor health. The AT1 transients lost nine of their twenty-two members within the next year, and no whale in this population ever had a successful birth again. As of 2013, there were only seven animals remaining in this population; it is only a matter of time before they become extinct. Killer whales in both these populations were seen swimming through the oil spill in the days after it happened; they likely experienced negative effects due to inhalation of toxic vapors. The transient whales may also have ingested the oil by consuming contaminated prey, such as harbor seals.

As part of NOAA's recovery efforts for Southern Residents, the administration finalized an oil spill response plan in 2009 to help keep killer whales away from contaminated areas using sound via helicopter hazing, banging pipes, and underwater firecrackers. The effectiveness of these methods in directing killer whale movements is known in part because similar techniques were used during the capture efforts of the 1960s and 1970s. While this may work to direct killer whales away from an oil spill, the long-term negative consequences of a large oil spill on the health of the Salish Sea are undeniable.

TOXINS: PERSISTENT ORGANIC POLLUTANTS

Persistent organic pollutants (or POPs) are aptly named. They're persistent because they're very slow to degrade. They're organic because they contain carbon, although they are almost exclusively synthetic, made either directly or as a by-product for industrial, agricultural, or medical uses. Finally, they're pollutants because they've been shown to have extensive health effects on both humans and wildlife. POPs are fat-soluble chemicals that bioaccumulate (the pollutant levels increase over time as an organism continues to consume toxic prey) and biomagnify (the higher an organism is on the food chain, the more amplified that organism's toxin levels are). The documented health effects of POP exposure are wide-ranging, including endocrine disruption, developmental defects, reproductive disruption, immunotoxicity, skeletal abnormalities, and increased occurrence of cardiovascular disease and cancer. Although primary exposure occurs for humans, killer whales, and other organisms through food, infant mammals have additional contact with these chemicals through their mother's fat-rich milk.

Three types of POPs are typically talked about relative to killer whales: dichlorodiphenyltrichloroethane (DDT), polychlorinated biphenyls (PCBs), and polybrominated diphenyl ethers (PBDEs). DDT, an insecticide developed during World War II, is one of the world's most well-known pollutants having been featured in Rachel Carson's *Silent Spring* in 1962. Widely used in agriculture, the most publicized effect is eggshell thinning in birds. Banned in the United States in 1972, DDT persists in the environment, though in declining levels. Widely used as coolants or heat transfer chemicals, PCBs were in hundreds of different products. Their main use was in transformers and capacitors, but they were also found in paints, dyes, fiberglass, and caulks. Production began in the 1930s and PCBs were banned in the United States and Canada in the late 1970s. The class of flame retardant chemicals known as PBDEs is perhaps the most widely used among the three POPs, as they are used in electronics, vehicles, furniture, building materials, clothing, and plastic toys. California was the first state in 2006 to begin banning manufacture and use of PBDEs. Washington and Maine have followed suit in the United States, and the European Union has banned two classes of PBDEs in electronics, but PBDEs are still widely used and their presence in the environment is rising.

These toxins enter the marine environment in several ways. Near heavily inhabited areas, contamination happens directly through runoff from agricultural and industrial applications. In the Salish Sea the Southern Residents are exposed to runoff and emissions from the more than seven million regional residents as well as at least ten pulp mills and municipal effluent from over twenty sources. But local sources of pollution aren't the only concern. POPs also enter the ecosystem indirectly, through atmospheric deposition. Chemicals from Asia literally ride the wind across the ocean, then essentially land on the water in the Pacific Northwest and become absorbed into the food chain. That means the two billion people from all around the Pacific Rim also impact the contaminant levels the Southern Residents and other marine species are exposed to. The global movement of chemicals in this manner explains why contamination occurs even in fairly remote areas, such as Alaska and the Arctic, although at lower levels.

In 2000, Peter Ross of the Institute of Ocean Sciences in Sidney, BC, and colleagues published a landmark paper in the *Marine Pollution Bulletin*. Using darts that took small skin and blubber samples, scientists collected forty-seven blubber biopsies from Northern Resident, Southern Resident, and transient killer whales in the mid-1990s. While the primary focus of the study was DNA analyses, it was also an opportune time to look at PCB contaminants in these whales. The results were surprising. The orcas of the Pacific Northwest are among the most contaminated marine mammals in the world, with all three populations showing higher levels than cetacean populations studied off the industrialized coasts of Europe and eastern North America. There was also a stark difference between the three killer whale communities and between the sexes within each community: transients are the most contaminated, followed by Southern Residents, then Northern Residents. Males are much more contaminated than females.

The differences in contamination levels of residents and transients is a result of their diet. Transients, which eat marine mammals, are essentially one step higher on the food chain than the fish-eating residents, so they are the recipients of an even greater level of biomagnified toxins, which gives them the dubious distinction of having the highest toxicity levels of any cetacean population on the planet. Meanwhile, Southern Residents show higher levels than Northern Residents because they spend time in more industrialized areas such as the Salish Sea and off the California coast. Both Northern and Southern Resident populations get the indirect exposure from chemicals crossing the ocean from Asia, but the Southern Residents get additional direct exposure by spending so much time in heavily populated areas. Studies of nonmigrating harbor seals in Puget Sound have shown that this area is a PCB hotspot. This indicates that while Southern Residents feed primarily on oceanic Chinook salmon, the species that make up smaller proportions of their diet—such as Puget Sound rockfish—may have a disproportionately large impact on the whales' toxin load.

The contrasting toxin concentrations between males and females is explained by their life history. Males essentially bioaccumulate toxins linearly throughout their lifetime. Females offload pollutants to their offspring during reproduction. Some of this happens during pregnancy, but most occurs during lactation, where calves receive an influx of toxins from their mother's fatty milk. From studies in other cetaceans, it has been estimated that female cetaceans offload as much as 60 percent of their toxin load to their calf. Reproductive females get the benefit of being less contaminated, but at the expense of pumping their offspring full of high levels of pollutants during key years of early development. This may contribute to the high mortality rate of young calves, particularly firstborns.

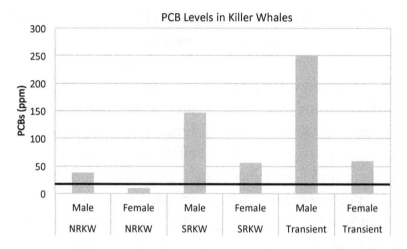

Average PCB levels (in parts per million) across three killer whale populations. *Blubber biopsy samples were taken from multiple killer whales of both genders across Northern Residents, Southern Residents, and transients. The horizontal line indicates the threshold for immunotoxicity in harbor seals. Data reillustrated from Ross et al. (2000)*

Interestingly, these patterns don't seem to hold up in precisely the same way when looking at PBDEs. For starters, PBDEs are still being produced and are increasing dramatically in the ecosystem, but they occur at several orders of magnitude lower than PCBs so concentration levels are measured in parts per billion instead of parts per million. Using the same set of samples from the mid-1990s, another research group looked at PBDE concentrations across the same three populations of killer whales. There was less of a difference between Southern Residents and transients, and no significant differences between males and females. The reason for this is not fully understood, but it could be due to such factors as increasing production levels of PBDEs (all whales may have fairly equal exposure to these newer chemicals, whereas older whales had much more exposure to old sources of PCBs), different environmental stability

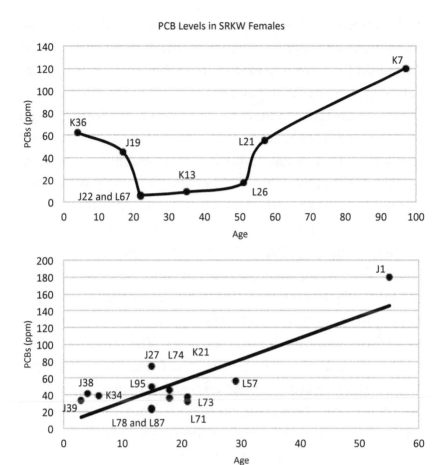

Trends by age in PCB levels (in parts per billion) across Southern Resident females (top, n=8) and males (bottom, n=13). *Males experience increased contaminant concentrations over the course of their lifetimes. Females experience a dip in concentrations during their reproductive years when they offload toxins to their offspring during pregnancy and lactation. Data reillustrated from Krahn et al. (2007) and Krahn et al. (2009).*

of these compounds, and/or different mechanisms of metabolic removal by the whales. For example, if PBDEs aren't offloaded by lactating females, there wouldn't be such a striking difference in concentrations between males and females. Essentially all members of the Southern Resident and transient killer whale populations have PCB concentrations above what has been experimentally determined as the threshold for immunotoxicity in harbor seals. But to what extent do the orcas experience negative side effects from these toxins? If the toxin concentrations alone were enough to influence health, we would expect based on observed levels that transients would be experiencing many more adverse health effects than either resident population. This is not what we're seeing, however, which indicates that the issue is more complex than simple levels of contamination.

According to NOAA's 2013 stock assessment for West Coast Transients, their population increased rapidly from the mid-1970s to the mid-1990s, coinciding with rebounding populations of some of their primary prey sources such as harbor seals and Steller sea lions. The transient killer whale population is still on the rise. Meanwhile, the 2013 stock assessment for Southern Residents explains that their population numbers also increased dramatically—about 35 percent—from the mid-1970s to the mid-1990s. But what came next was a population crash. Among the 20 percent population decline that occurred over five years were ten breeding-age males over age twenty. By the summer of 2001, there was just one breeding-age male in J-Pod, none in K-Pod, and three in L-Pod.

What is happening to the population, and to the males in particular? As is so often the case, there are more questions than answers. In March of 2000, just a few months before the Institute of Ocean Sciences's study on toxins was published, J18 Everett washed up in Boundary Bay near Vancouver, BC. The cause of death was determined to be a bacterial infection, but this particular strain often invades marine mammals who are already weak or sick, so Everett's situation was likely more complicated. It appeared the whale had not been eating for some time before his death. While we don't know for sure that toxins were the culprit, Everett's PCB levels were very high, and several hints showed that the pollutants could have been related to his poor health: his immune system was weak and his sperm count was zero.

Although POPs degrade slowly and thus persist for a long time in the environment, the good news is that banning them does have an impact. From 1972 to 1997 harbor seals in Puget Sound showed a 92 percent decrease in PCB levels. The bad news is that we haven't reined in PBDEs yet. Over a similar time period, from 1984 to 2003, PBDEs in Puget Sound harbor seals increased by an astonishing 6,950 percent.

NOT ENOUGH TO EAT

No fish, no blackfish. Those four simple words, a common regional saying, really get to the heart of the issue. These fish-eating whales rely on West Coast salmon runs, and preferentially feed on Chinook salmon, the largest and fattiest of the seven salmonid species. Pacific salmon declines have been a major environmental issue in the Pacific Northwest for well over a hundred years, and as the main prey item for Southern Resident killer whales, it's no surprise that the fate of the two species is inextricably linked. Salmon start their lives in freshwater streams and rivers, often in the forested hills and mountains deep in interior lands. As young juveniles, or smolt, they begin their migration to the ocean, traversing inland waterways until making the transition to the saltwater portion of their lives in coastal estuaries. They spend anywhere from one to four years, depending on the species, in the ocean, where they undergo long-distance migrations, traveling hundreds or even thousands of miles from where they entered the ocean. As full-sized adults, they reverse the entire process, swimming across the sea to find the entrance back to freshwater. Swimming upstream the entire way, wild Pacific salmon navigate back to their exact natal stream, returning to spawn in the very same stretch of water from which they emerged.

The life history of Pacific salmon is built on diversity, with each river and stream having its own unique stock with slightly different behavioral patterns, run timing, and genetic variation. Thus salmon are, by nature, resilient: well adapted to dealing with environmental changes, because even if one population can't adjust, there are likely some others nearby that can. Salmon have survived major climatic and environmental changes, including the last ice age and the resulting floods as glaciers receded that regularly rerouted regional waterways. They would be likely candidates to survive future challenges such as ocean acidification and global climate change as well—if given the chance. Wild Pacific salmon have been fighting against large-scale human impacts for more than 200 years. Salmon are the lifeblood of the Pacific Northwest ecosystem, a prey species for not only killer whales and humans but a wide variety of other animals; more than a hundred different species prey directly on salmon at different stages of their life cycle. In addition, salmon provide untold indirect ecological effects.

Just one example is the adults that die after breeding; their spawned-out carcasses decompose along stream banks and provide essential nutrients to the soil from the sea that help temperate forests thrive.

Coast Salish tribes had a sacred relationship with salmon, relying on them as a major food source, but also deeply honoring them through ceremonies such as at the return of the first salmon each year. Salmon's relationship with humans altered drastically with the arrival of Europeans to the Pacific Northwest in the 1800s. It didn't take long for white settlers to acquire a taste for salmon. Commercial fisheries exploded, bolstered by the fact that salmon are easy to capture, easy to preserve, and predictable in their returns each year.

As the newly arrived Euro-Americans did with many natural resources, they saw a valuable resource and proceeded to take as much of it as they could, with no regard for the future. Salmon canneries were filled with fish, many more than it was possible for the workers to can. In stark contrast to the Native residents of the Pacific Northwest, who traditionally used every fish they caught, the white owners/operators of the canneries felt it was preferable for salmon to rot, unused carcasses swept aside, than to halt production. By the 1840s the pressures of European fishing were impacting salmon populations, and by 1900 some stocks were starting to go extinct. One major problem with the new salmon fisheries was the fact that they fished in the open seas where salmon stocks from different rivers mix. Coast Salish fishermen typically harvested salmon as they returned to particular streams and rivers, which allows better management of individual river stocks, but the white

Members of J-Pod swim alongside a purse seiner in Haro Strait.

fishermen also took salmon from the ocean, which led to drastic overharvesting of salmon from certain areas.

Overfishing was detrimental to salmon stocks throughout the Salish Sea, but the human effects on fish didn't stop there. The construction of large, impassable dams blocked off entire tributaries that used to be home to numerous salmon runs. Even dams with fish passage caused high mortality with turbines killing outmigrating smolts and warm, slow-moving reservoirs behind dams proving detrimental or even fatal to salmon. Clear-cut logging damaged riparian zones, taking away the cool, shady waters so important to young salmon. Logging roads and debris further cut off some headwaters to spawning salmon, and more water was diverted from the rivers to support agriculture and a growing municipal population. Salmon hatcheries, which first started operating in the region in the late nineteenth century to help bolster salmon numbers for harvest, have outcompeted their wild cousins and diminished genetic diversity. Farmed Atlantic salmon, held in net pens until ready for human consumption, introduced an entire host of diseases to their wild counterparts. It didn't take long for salmon runs in all the major rivers feeding into the Pacific Northwest to plummet. Over the course of just a few decades, returns became a fraction of their historical levels. Wild Pacific salmon have experienced death by a thousand cuts.

Pacific salmon are one of the most studied groups of fishes in the world, reflecting their economic, ecological, recreational, and cultural importance both now and in the past. Despite the thousands of jobs, millions of dollars, and strong public support for salmon recovery, the returns on efforts have been minimal. Salmon recovery is a complex, polarizing issue with major trade-offs in short-term and long-term benefits. There are many divergent interests debating the pros and cons of various salmon recovery schemes. Advocacy groups represent commercial, recreational, and tribal fishermen; agricultural activities; forestry; electricity generators and users; natural resource management agencies; members of the environmental movement; endangered species and animal rights proponents; national, municipal, and local governments; and the general public.

With so many different groups at the table, the question of what salmon restoration even means is not a straightforward one. In their extremes, answers range from returning salmon to their pre-1850s levels to simply preventing species extinction. Does salmon recovery mean sustaining healthy wild fish populations with no hatcheries, or simply maintaining current harvest levels, even if that means via hatcheries entirely? The already hazy goals are blurred even further by conflicts of interest. Just one example is that the fish and wildlife agencies charged with regulating salmon fishing are supported in part by funds generated from the activities they have the ability to restrict. What's their incentive to curtail fishing when their budget is partially determined by the sale of fishing licenses?

The policy debate raises difficult questions that society has yet to ask, let alone answer. How much are people willing or able to pay for energy? Will fishing still be allowed, and if so, who gets first rights? How will public and private property be usable, and who gets to decide? Where are people allowed to live, and how much is the population allowed to grow? The question is not simply how important are wild salmon, but how important are wild salmon relative to other competing interests? Societal values change over time. In the mid-1900s, for example, such predators as cougars, wolves, sea lions, and bald eagles were seen as pests to be controlled or eliminated. Attitudes have shifted to not only tolerating these species but protecting them. The public's attitude about the vitality of salmon is shifting, from being yet another resource for people to exhaust to being key in their own right, a central part of the ecosystem and vital to life in the Pacific Northwest.

For salmon to adapt to a continually changing world, they need to have the genetic diversity within and between stocks. They need to have strongholds that act as refuges. Salmon will thrive, even in today's Pacific Northwest, if given half a chance. The lack of salmon is undoubtedly the biggest risk factor facing the Southern Residents, exacerbated by vessel threats and toxins. If there is a scarcity of fish, more ocean noise makes it harder for the whales to find what's there. If the whales aren't getting enough to eat, they are likely metabolizing more of the toxins stored in their blubber and experiencing the resulting developmental, reproductive, and immune system effects. If we make the oceans quiet and pollution-free, the whales still won't get enough to eat. If we increase salmon abundance, vessel effects and toxins are still problematic, but become less of an issue. The most action has been taken on vessel issues because that is the most visible problem and easiest to address from a management and policy perspective. Until the Pacific salmon runs these orcas rely on recover, however, neither will the whales.

LEARN MORE

For more on the role of lack of prey as a risk factor for the Southern Residents, see journal articles by scientists Brad Hanson, John Ford, and Graeme Ellis. For more on the risk of toxins, see the work of Margaret Krahn and Peter Ross. For more on the impact of vessel effects on the Southern Residents, see the publications of Christine Erbe, Juliana Houghton, Rob Williams, and Val and Scott Veirs. To learn more about the complex history of wild Pacific salmon management in the region, read *Salmon, People, and Place* by Jim Lichatowich and *The Salmon People* by Hugh W. McKervill.

CHAPTER 10

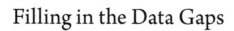

Filling in the Data Gaps

WITH THE ENDANGERED LISTING OF the Southern Residents, in addition to identifying key risk factors, the US and Canadian governments also recognized existing data gaps in our understanding of these whales. Despite having been studied for 30 years, we still had a lot to learn about factors that would influence our ability to help them recover. Topics for study included where all they traveled, what exactly they were eating and when, and how the risks they faced were influencing one another and resulting in impacts to the health of the whales. It's hard to imagine some eighty-odd killer whales disappearing for months on end, but many winters that's exactly what the Southern Residents do. They're tracked daily in the Salish Sea for many months during summer and fall, then one day they head out the Strait of Juan de Fuca to the open Pacific and may not be seen or heard from again until spring. The Southern Residents aren't migratory; they don't have a summer feeding ground and a winter breeding ground like the regional humpback and gray whales. Orcas are feeding and breeding year-round. For many years, when people asked where the Southern Residents went in the fall as they headed west out to sea, the best answer was, "We think they're on the outer coast, but we don't really know where." Once the Southern Residents were listed as endangered in 2005, however, answering this question became a priority.

WINTER RANGE, SATELLITE TAGGING, AND CRITICAL HABITAT DESIGNATION

Within the first year after the endangered listing, NOAA designated the critical habitat for the Southern Residents: An area that included nearly all of Washington's inland waters, with exclusions for several military sites. Similarly, the Department of Fisheries and Oceans designated most of the Canadian waters of the Salish Sea as Southern Resident critical habitat as required by their listing under the Species at Risk Act (SARA). Critical habitat designation is required for an endangered species, but in reality, it means very little. The only legal protection that results from a critical habitat designation is that new federal activities that are not already under permit within the critical habitat range must undergo a consultation (known as Section 7 in the US) before being approved. Under this process, if a federal activity is likely to adversely affect a listed species, a formal biological opinion on the impacts must be completed. Either alternative actions will be proposed or incidental take of the species will be allowed (basically okaying the activity). This does not apply to the excluded military zones, so the concrete benefits of a critical habitat designation are usually minimal. The greater benefit really comes from the increased awareness and education about which habitats are crucial to the listed species' survival. For instance, if other activities detrimental to the species occur within designated critical habitat, this fact can be used to lobby against them.

The original critical habitat designations pointedly left out the outer coast portion of the Southern Resident range, largely due to lack of data on where exactly they roam. The first oceanic sighting of the Southern Residents occurred when researcher Michael Bigg saw them off Tofino (the west coast of Vancouver Island) in 1982. The whales weren't spotted off the Washington coast until 1986. Between then and the early 2000s, there were only twenty total confirmed outer coast sightings of the Southern Residents. In 2004, with the endangered listing imminent, NOAA tried to increase outer coast reports. Multiple approaches were undertaken. Ken Balcomb and the Center for Whale Research provided education up and down the coast from British Columbia to California in hopes that more people would report sightings of the Southern Residents. NOAA conducted winter oceanic vessel cruises, during which they encountered the whales several more times. Finally, four hydrophones, and later seven, were deployed between the outer Strait of Juan de Fuca to Point Reyes, California, and passive acoustic monitoring techniques were used to detect killer whale vocalizations. By 2011 these efforts had collectively turned up more than one hundred additional winter outer coast reports of the Southern Residents. Scientists considered this relatively scant data, however, especially compared with the thousands of summer sightings of the Southern Residents in the Salish Sea.

Satellite tagging was the next logical step to determine more about the residents' winter travel patterns. A method that had been widely used on large cetaceans such as blue and humpback whales, NOAA's scientific review groups recommended that it now be applied to smaller species, including the endangered Southern Residents. In 2006, exactly thirty years after he had helped his professor drill pins through Flores and Pender's dorsal fins to attach radio-packs at the Seattle Aquarium, Brad Hanson again turned his attention toward the issue of how to tag a killer whale.

Tagging has come a long way since the 1970s, in part due to the work Hanson has done with new technologies to tag small cetaceans; his PhD work focused on tag development for porpoises. No longer is it necessary to capture the whales to surgically attach tags: smaller, dart-like tags can be deployed using a crossbow-style projector from a boat within thirty feet of the whales. The tags are secured to the whale by a pair of two-inch arrows that work their way out of the skin within three to twelve weeks. Whales can be tracked from further distances this way, as scientists with receivers don't have to be close enough to the tagged animal to pick up a radio signal. The newest transmitters, about the size of a 9-volt battery, send signals to NOAA's orbiting weather satellites, with daily sighting summaries emailed to scientists. The almost real-time sighting updates are available to a far wider audience. With Google maps, Twitter updates, and Facebook posts, it's possible for the word about the whereabouts of a tagged whale to spread quickly among scientists and the general public.

Starting in 2006, Hanson worked with other scientists to determine the impact of these new style tags on smaller cetaceans. One of his main goals was to confirm the tags were safe enough to use on small populations like the

Southern Residents. To monitor whale health after tagging, researchers collected data on wound repair, survivability, and fecundity of tagged animals. Nearly two hundred individuals of fifteen different species were tagged, and nowhere did they find negative impacts of tagging on the long-term health of any animal. In the short term, sometimes animals flinched during tagging but rarely was blood spotted. Other times, the animals didn't seem to react at all.

Various tagging studies showed how much this type of data could drastically increase and change what is known about the species being studied. For example, a tagged resident Alaskan killer whale traveled all the way to Kodiak Island, completely redefining the range for that population of whales. While researchers suspected the whales spent time at Kodiak Island, they had no idea how important of an area it was to that population, as was revealed by a few tagged animals. A blue whale tagged off Southern California was expected to travel south to the Costa Rica Dome. Instead, it went north as far as British Columbia. From 2010 through 2012 western gray whales off Kamchatka were tagged to determine where their wintering ground is, with many researchers speculating they would head down the Asian coast to the South China Sea. Instead, this group headed east across the Pacific and seemingly joined in with the eastern gray whale migration, traveling down the North American coast to Mexico. When it comes to predicting the movements of tagged animals, Hanson said that "almost always, preconceived notions are wrong."

In December 2011, with tag design and deployment methods refined and data showing low to no impact on the animal's health, the National Marine Fisheries Service (NMFS) granted a revision to Hanson's tagging permit allowing him to satellite tag up to six Southern Resident whales a year. As an additional precaution to protect breeding females and their young, he would only attempt to tag adult males and postreproductive females, no more than two whales per pod. Two months later, the first Southern Resident killer whale was satellite-tagged in the Strait of Juan de Fuca: J26 Mike. Unfortunately, just three days after deployment, the data stopped. The tag had apparently fallen off—a major disappointment, as researchers received on average a month's worth of tracking from other tagged killer whales.

Despite the legal permit, many members of the public were still against tagging Southern Residents. The short duration of the J26 tag didn't help. Concerned citizens believed the research technique was too invasive and posed too many risks regardless of the studies that had been done. Most agreed that the project goal—to determine more specifically winter habitats of the Southern Residents—was noble, but how many tags would it take for NOAA to officially designate any region as part of the whales' critical habitat? At a lecture about his research, Hanson deferred that question to Lynne Barre, a NOAA marine biologist of the Protected Resources Division who works on the policy side of the issue. Barre said there wasn't a specific number in mind and that they had to wait and see what kind of data were obtained. The more variable the data, the more data it would take. Her answer, an honest one, didn't satisfy many.

A NOAA research vessel prepares to dart a killer whale in Haro Strait. The person in the bow pulpit holds a crossbow, which is used both in satellite tagging and biopsy darting.

The only hard evidence of how much tagging it might take comes from other tagging efforts. NOAA had relied almost exclusively on twenty-three tagging events to determine the critical habitat of an endangered subspecies of false killer whales (another member species of the dolphin family but despite their name, not directly related to orcas) in Hawaii. For the Southern Residents, they've been seen from southeast Alaska to as far south as Monterey, California, but how they use this enormous range was still the unknown at the time of their endangered listing. We didn't know how far offshore they roamed, how often they went to California, where on the outer coast they might spend concentrated time, and, most crucially, what they're eating when they're out there. With the tagging of a second Southern Resident, K25 Scoter in December 2012, the first real data to help answer these questions started to emerge.

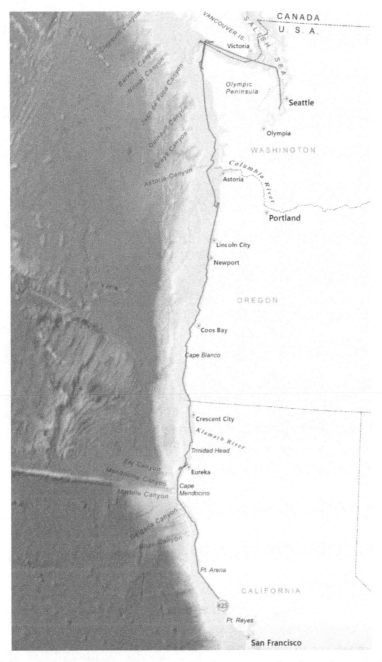

The first twelve days of Scoter's movements after tagging, when he and presumably the rest of K-Pod went all the way from Puget Sound to Point Reyes, California—a distance of more than nine hundred miles. NOAA.

L87 Onyx with a satellite tag visible on the left side of his dorsal fin. Photo by Sara Hysong-Shimazu.

One afternoon in January 2013, I pulled up a NOAA map showing the most recent movements of Scoter, who had been tagged in Puget Sound. After entering the open ocean, Scoter continued steadily south along the coast. A week after he was tagged, he was near Newport, Oregon, having traveled over four hundred miles. About five days later he had traveled south as far as Point Reyes—a total distance of more than nine hundred miles from where he was originally tagged. The latest update, two weeks after tagging, showed that Scoter had turned around and was heading north, following fairly closely along to the coastline.

Scoter's tag transmitted information to NOAA for ninety-three days. During this period Scoter and presumably the rest of K-Pod (at times also traveling with L-Pod) made multiple trips to California, going as far south as Point Reyes. They traveled from the Strait of Juan de Fuca to Point Reyes and back again in about three weeks. The biggest surprise was how close the whales stayed to shore, spending essentially all their time within five miles of the coast and rarely going out as far as the edge of the continental shelf (approximately ten to twenty miles offshore along the West Coast). As a result of this data, the hydrophones that passively detect Southern Residents were moved closer to shore for the following winter. On March 2, thanks to the information from the satellite tag, a NOAA research cruise was able to intercept K- and L-Pods off of Cape Blanco in southern Oregon. The scientists stayed with the whales for ten days; relocating them in the morning was a challenge but easier with the satellite tag data, compared with previous winter research cruise attempts where they relied solely on acoustic listening to stay with the whales overnight, a tactic that rarely worked. During this cruise, when the whales took them to Cape Flattery off the northern Washington coast, researchers were able to collect more than twenty prey scale or tissue samples and the same number of fecal samples. As expected, the

Southern Residents were eating Chinook salmon; more surprisingly, they were also taking steelhead, chum, ling-cod, and halibut in smaller amounts.

The data collected from these winter cruises and from the satellite tags are essential in expanding the Southern Residents' critical habitat designation to include their winter territory. "I feel a tremendous amount of responsibility under the government's Endangered Species Act contract to get these animals the protection they need," Hanson said. When asked about his critics, some of whom go as far as to say he is helping to kill the whales, Hanson said he fully expects his research to be criticized, which shows how much people care about the whales. Some people come at protecting the whales from a purely emotional standpoint, but his job is to come at it from a purely scientific direction, which means obtaining hard data.

Many anti-tagging people, including Ken Balcomb, have claimed in their public comments to NOAA that we already know all we need to about where the whales are in the winter. Hanson said NOAA will need hard data to back up Southern Resident habitat use, especially when it comes to holding up their endangered listing in a court case. He predicted that "this data [from K25's satellite tag] will come into play in some law suit in the future." One petition to de-list the Southern Residents came from a group of California farmers represented by the Pacific Legal Foundation in an attempt to claim additional water rights; it was denied by NOAA in 2013. "But they will be back," Hanson stated. "That's what they do." When the next petition comes, Hanson plans to have even more data on hand to defeat it.

Whale	Date Tagged	Date of Last Tag Transmission
J26 Mike	February 20, 2012	February 23, 2012
K25 Scoter	December 29, 2012	April 4, 2013
L88 Wavewalker	March 2, 2013	March 10, 2013
L87 Onyx	December 26, 2013	January 26, 2014
J27 Blackberry	December 28, 2014	February 15, 2015
L84 Nyssa	February 17, 2015	May 21, 2015
K33 Tika	December 31, 2015	February 17, 2016
L95 Nigel	February 24, 2016	February 26, 2016

Southern Resident killer whales satellite tagged under NOAA's effort to further define winter range of the population between 2012 and 2016.

It wasn't until July 8, 2013, that Scoter and the rest of K-Pod returned to the Salish Sea, more than six months after their last visit in late December, when the satellite tag was deployed. Photos of Scoter at the time showed that the tag had indeed detached, but that one of the attachment darts was still embedded in his dorsal fin. This came as a surprise to Hanson: after extensive tag redesign, he thought this type of detachment failure was impossible, especially in a cleanly applied tag like the "textbook" one on Scoter. However, due to precautions put into place specifically in the Southern Resident section of his tagging permit, the researchers ceased tagging until another redesign occurred.

The following year, L87 Onyx, who travels with J-Pod, was tagged with a newly remodeled tag. His travels showed that J-Pod made trips into Puget Sound, up and around Texada Island in the Strait of Georgia near Powell River, BC, and west out of the Strait of Juan de Fuca. Interestingly, they didn't go far beyond the western entrance to the strait before returning to inland waters, suggesting that J-Pod doesn't spend much time in the outer coastal waters, even in the winter months. This coincides with the fact that J-Pod is detected only on the northernmost of the seven outer coast hydrophones, and even then, infrequently.

Additional winter taggings, all of them on males, continued to reveal fairly consistent information: K- and L-Pods spend a lot of time on Washington's outer coast, particularly around the mouth of the Columbia River, and occasionally make trips either further south to California or into the Salish Sea. J-Pod, by contrast, remains almost exclusively in inland waters, rarely venturing more than a few miles from the western entrance of the Strait of Juan de Fuca. With this additional data, NOAA received a petition in 2014 from the Center for Biological Diversity to extend the critical habitat designation of Southern Residents to include the outer coast. While NOAA deliberated the petition, they continued to tag whales—until tragedy struck.

In the spring of 2016 a dead male killer whale washed up deep in an inlet on Vancouver Island. The body had decayed enough that it wasn't possible to identify the animal through his saddle patch markings, but the presence of wounds from a satellite tag led to an eventual identification of the whale as L95 Nigel, who had been tagged over a month prior. The general public jumped to the conclusion that the tag killed the whale, but facts about the incident remained muddy for months. Eight months later, the necropsy report was released, piecing together what had happened. The cause of death was listed as a fungal infection, with the infection occurring at the site of the tag wound. The report revealed that the tag was not properly sanitized after an initial tagging attempt had failed. More than five hundred satellite tags had been deployed on more than a dozen species up to this point, and the first fatality occurred on a member of an endangered population. NOAA suspended indefinitely the satellite tagging program of Southern Residents, while delaying a final ruling on a critical habitat extension. "We knew there was a risk. We felt it was fairly small," Hanson said. "It's not been easy. It won't be. It's something I'll carry with me for the rest of my life."

PHOTOGRAMMETRY AND ASSESSING BODY CONDITION

Photogrammetry is the process of obtaining measurements from photographs, with applications that range from topographic mapping and meteorology to engineering and biology. When the distance from the camera to the object of interest is known, the scale of the image can be calculated and actual measurements of objects in the photos can be obtained. In the early application of photogrammetry to killer whales in the 1980s, researchers such as Jim Boran didn't have distance measurements, so they resorted to using ratios that were measurable within a photograph. For instance, Boran estimated calf growth rate by comparing the relative size of a calf to its mother from year to year. He looked at dorsal fin height versus width to get a sense of growth rate over time. Several decades later, Boran pointed out the shortcomings of his research. The biggest problem was no ability to account for the angle of the whale relative to the camera: unless the animal was perpendicular to the camera, all the measured ratios would be off, and as he knew from looking at thousands of images of killer whales, a slight difference in angle could make a huge difference. "Even then I knew overhead images were the way to go for photogrammetry," Boran said. "But I worked with what I had, and still, it was a start."

The next advances in photogrammetry on killer whales didn't come until more than twenty years later, in 2004. While accompanying an Orca Survey photo expedition in Puget Sound, NOAA researcher John Durban tested out a new method of measuring dorsal fins. By fitting a custom bracket to the top of a Nikon camera, he and his partners fitted two laser pointers mounted ten centimeters (about four inches) apart to the camera lens. The lasers had a range of one hundred meters and could be projected onto a whale's dorsal fin, providing something of known scale on the image so that actual dorsal fin heights could be measured. This methodology is called lasermetrics. (Boran had also attempted to get something of known size to help him attain measurements in his photographs: he had a friend throw a Frisbee toward surfacing whales while Boran took photos. It didn't work very well.) Durban's results confirmed that there can be considerable individual variation in growth rates. The dorsal fins of J26 Mike and J27 Blackberry, two young males both born in the same year, were more than three inches different in height.

In 2008, Durban was back to look at the Southern Residents with the NOAA Southwest Fisheries Science Center's photogrammetry team, this time from the air. Using unique cameras originally designed for reconnaissance missions for the US military, this group of researchers specialized in taking high-resolution aerial images from a helicopter to survey marine mammal populations. Having previous experience with whales and dolphins as well as seals and sea lions, the team used the cameras to monitor body size and estimate group sizes—for example, the number of hauled-out seals. It turns out that killer whales are excellent animals to study via aerial photogrammetry, because their unique black-and-white coloration give several "landmarks" to help with accurate measurements. The eyepatches help make the animal's health very apparent even before exact measurements are taken. From above, healthy whales have eyepatches that appear to bulge away from their body, indicating a robust animal. Skinny whales, by contrast, have eyepatches that appear to go straight back, or even curve in toward the body, when there's no fat buildup behind the skull.

The photogrammetry team completed ten successful helicopter flights over two-and-half weeks and obtained nearly three thousand images of sixty-nine different whales. They focused their efforts on forty-six whales for which they had more than five high-quality images and thus could calculate averages. The smallest animal at the time was the calf K42 Kelp, who was less than a year old. He measured just under nine feet in length. Adult females ranged in size from eighteen feet (J22 Oreo, the shortest) to just over twenty-one feet (K14 Lea, the longest). Adult males ranged in size from just over twenty-one feet (K21 Cappuccino, the shortest) to just under twenty-four feet (L41 Mega, the longest). On average, Southern Resident females were 19.72 feet long and males were 22.12 feet. While Lea and Cappuccino were within a few inches of each other in terms of overall length, there was no overlap between adult males and females among any of the whales measured. Just how accurate were these measurements?

To figure out the error in calculations, on each flight the helicopter also took images of one of the Center for Whale Research boats, which were of known lengths. By comparing their calculated lengths from overhead images of the boat to the actual boat length, they figured out their measurements were accurate to within three inches.

In addition to looking at whale lengths, Durban was interested in taking other measurements that might be indicators of body condition. Using the same set of images, the researchers measured breadth of the whale right behind the dorsal fin and head width right behind the cranium. There was more variability between whales for head width, so the team focused on this characteristic as a potential indicator of body condition. They calculated averages from five or more photographs for thirty whales, and came up with a ratio of head width to body length expressed as a percentage. The largest proportional head width was that of the calf Kelp, measuring at 18 percent of his body length. The smallest head width was that of his mother, K14 Lea, at 10 percent, perhaps indicating the energetic toll nursing a young calf can take on a new mother.

However, the proportional head width measurement failed to capture the fact that L67 Splash was in poor body condition at the time she was photographed in mid-September. The boat-based observations determined she was thin and had a peanut-head depression behind her blowhole indicating malnourishment. This encounter turned out to be the last time she was seen, as she was presumed dead shortly thereafter. When looking closer at their data, researchers found that Splash had the second smallest actual head width to Lea; when they calculated the proportions based on body length, Splash's measurements did not stand out because she was a shorter-than-average female at less than nineteen feet long. Coming up with a threshold—for instance, saying that a whale with a head width proportion of less than 12 percent was in poor health—was going to be difficult because of the amount of variability between whales. This sentiment echoed what Dave Ellifrit of the Center for Whale Research said: you have to look at each whale's personal history to assess its current health. The best way to assess body condition is to photograph the same individuals across different years, to see how their bodies change over time. Photogrammetry is a perfect tool for studying killer whale health, because it can determine how animals are doing without waiting for one to die to reveal a problem. With the Southern Residents, where all whales are known and can be monitored, photogrammetry actually monitors the growth and health of specific individuals over time.

In September 2013 the same photogrammetry team reassembled on San Juan Island. By the end of the month, they had accumulated nearly seven thousand aerial photographs of sixty-nine different whales, including more than fifty whales that were documented during the helicopter flights five years earlier. This allowed the researchers to look at changes in body condition for the same animals by comparing head widths from photos taken in 2008 to those taken in 2013. Of the whales measured in both years, sixteen showed significant changes in head width; eleven of these were declines, while only five were increases, providing support for the theory that the whales were going through a difficult salmon year in 2013. Overall, proportional head width decreased across all age/sex classes except adult males. This led Durban to speculate that maybe females, who are known to prey-share, may be suffering in part because they're sharing limited food to help keep the males healthy. (This ties in to the previously mentioned fact that males are fourteen times more likely to die after their mothers do; they may be relying on them for food.)

Researchers developed a method to detect pregnancy in reproductive-aged females from aerial photographs. By measuring whales who were known to be pregnant at SeaWorld San Diego, researchers discovered that pregnancy could be detected by a whale's girth behind the dorsal fin as early as four to six months into the pregnancy, with increased girth becoming very apparent by thirteen months. With this knowledge, researchers detected twelve potential pregnancies among the Southern Residents. However, only two of these females produced a known calf, suggesting a high reproductive failure rate. Two births from twelve pregnancies is a very low ratio, but scientists won't know how abnormal this is until they're able to conduct comparative studies with other, healthier killer whale populations—namely the Northern Residents and transients, who both have steadily increasing numbers. Killer whales may have a high rate of reproductive failure in general, so researchers won't know how the Southern Residents measure up until other populations are studied.

A handful of years ago, precise photogrammetric measurements from a helicopter was the cutting-edge technology, but today a new methodology has been developed using unmanned aerial vehicles (UAVs), also known as drones.

Durban and the photogrammetry team built a custom hexacopter for studying marine mammals from the air. Just three
feet across with six small motors and operated remotely
from a boat, the drone provides a quiet, stable, maneuver-
able, and portable platform for conducting photogrammetric studies. Flying it is much cheaper than renting and fuel-
ing a helicopter. The hexacopter flies about a hundred feet
above the whales, and the resulting images are much clearer
than those taken at nine hundred feet from a helicopter.

The photogrammetry team—researchers from National Oceanic and Atmospheric
Administration (NOAA), SeaLife Response, Rehabilitation, and Research (SR3),
and Ocean Wise/Vancouver Aquarium—launches a research drone near San
Juan Island.

L121 Windsong nurses from his mother, L94 Calypso, in 2015. Image captured by drone flown by NOAA, SR3, and Vancouver Aquarium researchers.

After conducting pilot study flights on the Northern Residents, the photogrammetry team not only had the beginnings of a strong dataset on this population but had also determined several other applications for their ability to take aerial photographs of the whales. When Northern Resident I105 got entangled in a fishing net, for example, the drone was able to confirm after the whale was freed that it wasn't trailing any additional gear. Another Northern Resident, A95, was injured by a boat propeller, and the photogrammetry team was able to check on the status of her injuries as well as document her remarkably quick healing. In addition to aiding in entanglement and boat collision response, the hexacopter documented just how often killer whales like to swim upside down. As a result, the researchers determined genders for more than twenty whales for which gender was previously unknown. They can also see salmon in their photographs and thus determine the species and measure the size of the fish the whales are pursuing, catching, and sharing. Additional applications will become apparent as time goes on, as researchers monitor changes in things like whale spacing from one another and their skin conditions. Durban and his team flew test flights at one hundred, sixty, and thirty feet over Northern Resident whales but never saw a behavioral response from the animals, confirming that this is a noninvasive research method. That doesn't mean anyone can fly drones over the whales, however; permits are required, and the two-hundred-yard no-approach rule for vessels in the United States applies to drones as well. Citations of more than one thousand dollars have been handed out.

After just one week using the hexacopter on the Southern Residents, it was clear to Durban that the drone results in a lot more, higher-quality data than the helicopter. The plan going forward is to conduct annual September flights on the Southern Residents, building up the dataset to study how individual animals are growing and changing over time. Ideally, he wants to have both spring and fall flights so the team can also look at changing body condition throughout a single season. While the photographs that result from the studies are stunning in their own right, the hope is this work will demonstrate more directly how the whales' health relates to salmon abundance in the region. So far, from 2008-2018, their work has shown a decline in body condition across the Southern Resident population, with the worst declines in J-Pod. "The data are consistent with nutritional stress," John Durban said in a lecture in fall 2018. "It is a no brainer that they need food."

PICKY EATERS

Identifying salmon that whales chase and eat via hexa-copter surveys may provide new insights into killer whale foraging going forward, but researchers have used many methods over the years to determine what the whales are eating. Detailed field studies on orcas began in the Pacific Northwest in the 1970s and elsewhere around the world shortly thereafter. Although killer whales have adapted to eat almost everything in the ocean, each pop-ulation specializes on one or a few specific prey types that occupy their region. The list of what killer whales eat as a species is long, but the list of what an individual orca eats is quite short.

The dietary preferences of killer whales seem to be the evolutionary driver that leads to so many different popu-lations around the world, as certain whales refine special hunting methods for their preferred prey and share these techniques with their families. Scientists determined this through careful observation, and except in some cases where stomach contents were analyzed, they learned this by watching what killer whales hunt and eat, sometimes collecting scraps like fish scales at the surface to help identify prey species. Researchers began to discern the difference in diet between residents and transients in the

K25 Scoter surfaces with a salmon in his mouth.

1980s, but it wasn't until the late 1990s, when John Ford and Graeme Ellis published their summary of twenty years of field observation, that residents' preference for Chinook was firmly established. In the mid-2000s, Ford and Ellis did another dedicated field study on resident killer whale feeding behavior, which determined that 95 percent of resident orcas' feeding events were on salmonid fishes. They also presented the first solid evidence for prey sharing among residents. About three-quarters of the time they were able to confirm a whale had caught a fish, two or more other residents would converge and circle for a couple of surfacings. Their observations suggested that while resident orcas pursue and hunt salmon individually, the whales regularly share their catch with immediate family members, with female residents being the most likely to share with their offspring.

The emerging picture was of killer whales, particularly residents, being picky eaters, but the data from Ford and Ellis amounted to only a few hundred observations over thirty years, and just a few dozen of these were from the Southern Residents (most were from the Northern Residents). When the Southern Residents were listed as endangered in 2005, the scientific community felt an urgency to gain a better understanding of which specific salmon stocks these whales relied on. If the Southern Residents were going to benefit from protection of a vital prey source along with their endangered listing, more data was necessary, and Brad Hanson set out to get it.

Over five years Hanson and his research team, primarily through focal following individual whales (pursuing them with a federally issued research permit that allows close approaches), collected more than three hundred scale, tissue, and fecal samples from Southern Residents' predation events. Of these, 227 samples contained enough material to identify the species preyed upon: 78 percent contained Chinook salmon. Also present in the samples, particularly in the spring and fall, were steelhead. Noticeable for their absence were pink and sockeye salmon, which are up to one hundred times more abundant in the region than Chinook. This supported Ford and Ellis's claim that the Southern Residents, like the Northerns, were preferentially feeding on Chinook over other more abundant salmon species. Here at last was the substantial data showing what was long suspected: the Southern Residents, like their Northern counterparts, preferentially fed on Chinook. But which Chinook?

To quote the old adage, there are a lot of fish in the sea. With so many rivers and streams hosting Chinook salmon populations along the West Coast, which ones are of particular importance to the Southern Residents? In the mid-2000s, coinciding with Hanson's work, a new project called the Genetic Analysis of Pacific Salmonids was developed. Ten laboratories ranging from the Gulf of Alaska to central California pooled their resources to share genetic samples of more than 150 Chinook salmon stocks from the West Coast. By extracting DNA from Hanson's prey samples, geneticists compared the sampled salmon to this new database and were able to determine which stocks the salmon from the predation events belonged to. The results were striking: About 80 percent to 90 percent of the Chinook the Southern Residents had eaten during the summer months came from the Fraser River, which meets the Strait of Georgia right at the city of Vancouver. Which Fraser River stocks they were feeding on in particular depended on the season, moving from the Upper Fraser to the Middle Fraser to the Lower Fraser as the summer progressed. Puget Sound salmon stocks, comprising about 15 percent of Chinook that return to inland waters, made up a similar percentage of the killer whales' diet.

Hanson's work eliminated any questions about what the Southern Residents ate in summer, but their winter diet was largely unknown. To answer this question, Hanson got on the water with the Southern Residents as they made visits into Puget Sound between October and January. Although he collected fewer overall samples than he had in the summer months, Hanson's results showed that Chinook were still an important part of the diet, but consumption of chum salmon also rose dramatically. In fact, both species were consumed in about equal proportions. But what did the resident whales consume when they were along the outer coast? Beginning in 2004, and continuing most winters since then depending on funding and vessel availability, Hanson's team at NOAA has taken one- to three-week cruises on the outer coast. They've almost always encountered the Southern Residents, with years during which satellite tagging occurred drastically increasing the time spent with the whales. When weather permitted, the team deployed smaller research boats in an attempt to gather prey and fecal samples.

The first few samples collected in March 2009 along the outer coast were of Columbia River Chinook; a few dozen samples taken in the following years showed that Chinook and chum remained a part of their outer coast diet throughout the winter, but other species like halibut, lingcod, and steelhead have also shown up. Hanson suspects steelhead in particular may prove to be an important winter prey species. Sample collection has continued during every outer coast survey, but many of these samples are sitting in a freezer at NOAA headquarters in Seattle. The delay in analyses, attributed to lack of funding, means that important information about what fish stocks the Southern Residents are relying on during the winter is tantalizingly close but still relatively unknown.

INTERPLAY AMONG THE RISK FACTORS

While all three identified risk factors—limited prey, toxins, and vessel threats—play a role in the decline of the Southern Residents, toxins and vessel concerns become more of an issue when there aren't enough fish for the whales to eat. For instance, toxins stored in the blubber become more serious health risks when fat stores are metabolized because a whale isn't getting enough to eat, and boat noise that may influence the ability to detect prey becomes more of a problem when there are fewer fish to find in the first place. The Center for Conservation Biology at the University of Washington came up with a unique solution to tease apart the relationship between these risk factors while learning more about the specific stocks the Southern Residents eat. The researchers run a program called Conservation Canines, where scat detection dogs aid in field research by detecting fecal samples from the study species. The program uses similar methods that law enforcement officials use to train narcotics detection dogs. The dogs selected for the program, all rescue animals, must have exceptional

Black lab Tucker searches for the scent of whale feces from the bow of the Conservation Canine research vessel.

focus. The dogs' reward when they detect a sample is that they get to play with their ball. The idea of using dogs for field research was appealing, not only because of the noninvasive method of sampling an animal's feces rather than the animal itself, but also because it allowed new insights into otherwise difficult to study species that are far-ranging, elusive, nocturnal, or unapproachable. The program had successfully surveyed such species as grizzly bears, spotted owls, and wolverines. In September 2006 they ran a pilot study to see if dogs could help researchers locate fecal samples from whales.

Obviously, using dogs to study killer whales presented challenges—namely the need to navigate on water rather than on land and the fact that whale feces sinks or dilutes into the surrounding water rather quickly. The inaugural attempt fell to graduate student Katherine Ayres and a new Conservation Canine recruit, an energetic black lab named Tucker. Ayres and Tucker helped set up a procedure used by future graduate students and a few other dogs, although Tucker remained an ace in the whale feces department until his retirement in 2017. Because they're on a boat, the dog has to lead the handler to the sample through body language. The boat driver essentially has to zigzag according to signals interpreted by Tucker's handler, moving the research team through the "scent cone" of the smell until the humans are close enough to detect and scoop the feces out of the water. Handlers have determined that Tucker can locate a fecal sample from up to a nautical mile downwind of where it's located.

Starting in 2007, the Conservation Canine team spent summer-long field seasons in the San Juan Islands collecting data. Their research boat became a mainstay on the water and provided an excellent opportunity to educate the public about killer whale research and important conservation issues. Spotting "the pooper scoopers," as they are affectionately known among the whale-watch community, is a highlight of many trips. Members of the public love seeing the dog in action, laugh about the idea of whale poop, and are intrigued by the noninvasive research technique (the boat often collects data a long distance from the whales). The data collected led to Ayres's PhD dissertation and numerous publications as a result of this novel research technique.

The fecal samples collected by the Conservation Canine team are shared with NOAA's Northwest Fisheries Science Center, where Hanson's research team tests the samples to determine which prey species the whale has eaten. A 2016 study by NOAA's Mike Ford and colleagues used DNA of prey in whale fecal samples to make another assessment of the Southern Residents' summer diet. This had the potential to reveal a more complete picture of the whales' overall diet since all species eaten will be represented in fecal matter, whereas the diet determined by scale collection at the surface may be biased toward prey consumed near the top of the water column where bits of prey are likely to float to the surface for researchers to collect. Like Hanson's direct scale-sampling study, the fecal matter diet assessment found that Chinook made up about 80 percent of the Southern Resident killer whale diet. Interestingly, they also showed that coho salmon were being consumed in late summer, indicating this salmonid species may be a more important component of the diet than previously realized. Sockeye were also detected in a small percentage of the samples.

Rather than DNA analyses, the research of the Center for Conservation Biology, however, focuses more on conservation physiology, a relatively new field that looks at the physiological responses of organisms to human alterations of their environment. Fecal material contains a lot more information than just what a whale has eaten—it turns out orca poop is a biological gold mine. Scientists can collect DNA from individual whales via their fecal matter as well as hormone levels of a particular whale at a particular time. Hormones are released in the body in pulses, but in a fecal sample hormones capture what's happened in the body over a longer period of time—say, the entire digestive period. This gives researchers a broader picture about the conditions the whale has experienced.

The fecal thyroid hormone T3 was one of the elements examined in a monthlong pilot study. Thyroid hormones are related to nutrition (levels increase the more a whale eats), and they are also affected by toxins. Preliminary results showed a difference in thyroid levels between different pods. L-Pod had the lowest thyroid levels, K-Pod had moderate levels, and J-Pod had the highest levels. A statistical analysis showed significantly different levels between J- and L-Pods, but why? Perhaps J-Pod had access to better or higher-quality prey. Another theory is that L-Pod whales, which travel further, metabolize more of their blubber reserves and as a result release more toxins that interact negatively with thyroid levels. In any case, this data collection showed a glimpse into an entirely new research field. A lot can be learned from the scat collection dogs.

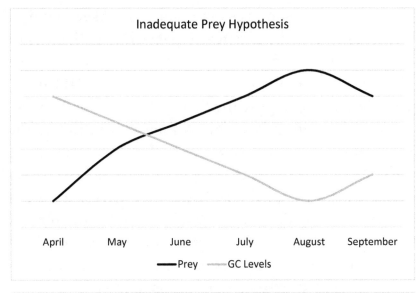

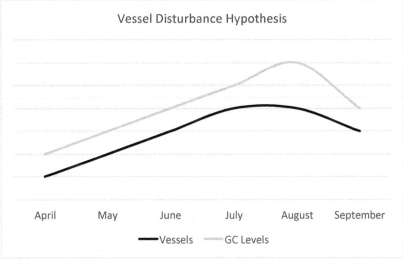

A visualization of the "inadequate prey" versus the "vessel disturbance" hypothesis using a fictitious data set.

Ayres's PhD research focused on two hormones: fecal thyroid (T3) and glucocorticoid (GC). GC levels increase in response to nutritional and psychological stress, while T3 declines in response to nutritional stress but does not respond much to psychological stress. Ayres's goal was to tease apart whether the whales were affected more by inadequate prey levels or the close proximity of vessels, or by both potential threats equally. During August and September the whales experienced peaks in each factor: salmon availability, as the summer Chinook stocks return to the Fraser River, and vessel traffic, as this is the height of the whale-watching season. If prey availability is the main stressor on the whales (the "inadequate prey hypothesis"), the prediction is that GC levels will be at their lowest during these prime salmon runs. If vessels are the primary stressor (the "vessel disturbance hypothesis"), the prediction is that GC levels will be higher during this time, when the whales are exposed to the most number of boats on a daily basis. If both salmon and boats are affecting the whales, the expectation is to see a stronger correlation between GC levels and boats during years of lower salmon abundance. In addition, since T3 responds primarily to nutritional stress, the prediction is that T3 would only correlate with vessel numbers if the presence of boats was impacting the whales' ability to forage.

The results of Ayres's study showed the same trend across three years: GC levels, indicating stress, were at their lowest when the Fraser River Chinook salmon runs peak at the end of August each year, supporting the inadequate prey hypothesis. This is when the whales were the least stressed, despite the fact that this is also the time of year when they are experiencing the largest number of boats in their vicinity. Somewhat surprisingly, the T3 levels measured in the whales' scat samples were at their highest when the pods returned to inland waters in the late spring or early summer. Since T3 levels correlate with nutritional status, this indicates that whatever the whales are feeding on, presumably along the outer coast, during the spring months plays a more important role in their survival than was realized. The summer Fraser River Chinook salmon runs may not be the biggest feast the Southern Residents enjoy during a given calendar year. Over the course of the summer, T3 levels steadily declined from their spring highs until mid-August, when they rose slightly, in correlation with the peak of the summer salmon runs. Conversely, T3 levels did not statistically correlate with vessel abundance numbers. Ayres's study concluded that the whales' nutritional status correlates with salmon abundance and not with vessel abundance, strongly supporting the hypothesis that lack of prey—not vessel traffic— is the key factor affecting whale stress.

Researchers have come a long way in understanding the dietary preferences of killer whales: Southern Residents eat fish, predominantly salmon, with a preference for Chinook, specifically Fraser River Chinook in the summer. Resident killer whale survival is strongly linked with coastwide Chinook salmon abundance. Although there aren't any links with particular river systems, the numbers show that when there are fewer Chinook salmon in the North Pacific, more Southern Resident killer whales die. In some cases, the orcas may literally be starving, but lack of food complicates their health in other respects. When they're not getting enough to eat, they're also more susceptible to disease and less likely to give birth. Cumulatively, what the science has shown is undeniable: lack of prey is the biggest hurdle in the way of recovery of the Southern Resident killer whales.

LEARN MORE

Detailed information on the Southern Resident killer whale satellite tagging program, including maps of data from tagged animals, can be found on NOAA's Northwest Fisheries Science Center website. For more on modern photogrammetry work, see the research publications of John Durban and Holly Fearnbach. Work on prey studies has been published by John Ford, Graeme Ellis, and Brad Hanson. Publications by Katherine Ayres and Sam Wasser go into more detail on the work of the Conservation Canines using fecal samples.

CHAPTER 11

Asking the Tough Questions

WHERE ARE THE WHALES? IT'S a question whale-watch naturalists patiently put up with from May through September, the best time of year to see the Southern Residents in the Salish Sea. Second only to "What time do the whales come by?" the question "Where are the whales?" is a common one not only from tourists but also among researchers, boat captains, naturalists, and all members of the whale-watch community. In the summer the answer to the whales' whereabouts is usually something like "On the west side of San Juan Island" or "Swimming past Victoria" or "Near the mouth of the Fraser River." In 2013, however, the question took on a different meaning among the whale-watch community of the Salish Sea. It became a sadder, more anxious question because that summer, quite simply, the whales weren't here.

WHY AREN'T THEY HERE?

It's true that whale-watchers saw all of the members of all three pods (J-, K-, and L-) in inland waters that season, but among those of us who watch the Southern Residents year after year, the consensus was that something different was definitely going on in 2013. They came back later than usual. Their visits to Haro Strait and the Strait of Georgia were much shorter than usual. For a couple of bizarre weeks the only Southern Residents in the Salish Sea were the three members of the L22 matriline—something that had never been witnessed before. April and May used to be considered good whale-watching months in the Salish Sea. According to long-term sightings records, from 1976 to 2008, members of at least one of the three pods of Southern Residents were documented in inland waters in every single month of every single year. During the winter season this may have been for just a day or two each month, but most years J-Pod was around nearly every day in the spring.

In 2009, for the first time on record, there were no sightings of Southern Residents anywhere in the Salish Sea in the month of April. The same thing happened again in April 2013, except this time the trend persisted all summer. Researcher Bob Otis, who tracks "whale days" at Lime Kiln Lighthouse during the summer, reported an average of more than forty-five days of whale encounters per year during his May–August study season from 1990 to 2012. In 2013, though, the whales were around just eighteen days during the same time period. Why were they spending less time here? It all comes down to food. The Southern Residents have shown that they will travel far and wide to look for fish if they're not finding it in their traditional core summer habitat of the Salish Sea. After the spring Fraser River salmon took a crash in the mid-2000s, the orcas have spent less time here during April through

June. May 2018 became the first May on record with no Southern Residents in inland waters.

During the worst years of the Fraser River salmon runs, which are sadly becoming more frequent, the absence of the resident killer whales extends into the summer months as well. This happened in 2013, and to an even greater extent in 2017—as Chinook salmon tend to spawn in four-year cycles, this wasn't a coincidence. Where the whales are going during these low salmon years is currently unknown. During spring they may be taking advantage of spring Chinook runs returning to the Columbia–Snake River Basin, as on a year-to-year basis these runs may be comparatively better than Fraser River runs. In summer they may be spending more time on the outer coast of British Columbia. The fact that the whales' habitat is shifting, however, should provide further impetus to expand their critical habitat. Unfortunately for whale-watchers in the Salish Sea, what was traditionally the daily feeding grounds of the Southern Residents have become less so in recent years.

The timing of the Southern Residents becoming less frequently observed in the Salish Sea has coincided with an increase in humpback whale and transient killer whale sightings in regional waters, indicating that at least parts of the inland water ecosystem are healthy. The trends between resident killer whale absence and transient killer whale presence are probably not entirely unrelated. The decline of salmon has coincided with an increase in transient killer whale prey—including harbor seals in the Salish Sea. Harbor seals eat juvenile salmonids, which prevents the fish from growing up and coming back as resident whale–sized prey.

Some people blame harbor seals for salmon declines, but it's not that simple. Food webs are complex. In most locations, salmon are only a small portion of harbor seal diet, while they also eat other predators of juvenile salmon such as hake. In addition, anecdotal observations have shown that transient killer whales avoid resident killer whales, in that they often change directions to avoid direct interaction with residents. For transients, the combination of fewer residents and more seals has likely made the Salish Sea a more attractive place to hang out. In the summer of 2017, the Southern Residents were around on fewer days than ever before recorded, and simultaneously transient sightings were off the charts.

It's been interesting to see how whale-watchers—both people and companies—have reacted to the Southern Residents being around the inland waters less often. On the one hand, whale-watch companies still see whales, including orcas, on a high percentage of their trips. For people who don't know the difference, it's not apparent that the Js, Ks, and Ls of the Southern Resident community are no longer spending time in the inland waters. On the other hand, some people are so keen to see specifically the Southern Residents that they are disappointed by anything else. One local whale-watch company had a passenger review that said: "We did see transient orcas. Better than nothing." It raises a lot of questions that the region's naturalists and educators are discussing. How do we convey the message of the plight of the Southern Residents without being too depressing? How do we get people excited about all the whales and wildlife they *do* see? How do we make sure the story of the Southern Residents is still told, and they don't simply become forgotten?

The Salish Sea was identified as the core summer habitat for the Southern Residents and as such was given critical habitat designation by both the United States and Canada. With more data from the winter satellite tagging efforts, the Center for Biological Diversity petitioned NOAA in 2014 to extend the critical habitat to include the outer coastal waters. With the Southern Residents now spending much less time in the Salish Sea than they have historically, extending the critical habitat designation is more important than ever, yet NOAA has not made the expanded designation. In 2015, a year after receiving the petition, NOAA released a document that essentially said they would continue to review the request for the next two years, delaying a decision. 2017 came and went without a final decision, so as of 2018 the Center for Biological Diversity filed a legal notice warning the federal government that their continued delay is illegal and a lawsuit will follow if no action is taken by NOAA. Meanwhile, in late 2018, Canada did extend Southern Resident critical habitat in British Columbia to include some of the outer coastal waters off southwest Vancouver Island.

WHY AREN'T THEY HAVING CALVES?

Within hours of its discovery in December of 2014, the news of a stranded killer whale on Vancouver Island's Bates Beach near Comox, British Columbia, went viral among Facebook's whale community. Rumors ran rampant: it's a transient, not believed to be a resident, there's no match in the transient identification guides, it's a subadult male, it's a female. I didn't know what to believe. A photo was posted and the rumors turned to educated speculation after the saddle patch and dorsal fin were seen. I felt sick when researcher Ken Balcomb confirmed that the deceased whale was eighteen-year-old J32 Rhapsody. A large group of whales, presumably J-Pod, had been seen near Comox on December 3. When Rhapsody was first seen floating off the beach on the morning of December 4, she had probably been dead less than twenty-four hours. It was a surreal experience to watch the play-by-play of discovery, identification, photos, and rumors of cause of death play out minute by minute on social media.

It's sad to lose any whale, but particularly so when it's a breeding-age female, the age/sex class so critical if the Southern Resident population is to have a chance at recovery. Earlier in the season, rumors spread that Rhapsody might be pregnant, because she looked particularly robust in several breach photos. She was always a round whale, and there are few reliable visible signs of pregnancy on orcas "(at least from the side - as discussed in the photogrammetry section, pregnancy can be reliably detected from above)", so we didn't know if she was really carrying a calf or not. At eighteen years of age, we all surely hoped she was. The silver lining in Rhapsody's death would be the knowledge we could gain from her. Was she fertile or infertile? Was she pregnant, or has she ever miscarried? What was in her stomach? What were her toxin loads? What diseases did she carry? Why did she die?

A necropsy occurred two days later with dozens of people, including Balcomb, in attendance. The first news to emerge that day was that several of Rhapsody's teeth had been illegally sawed off overnight and taken as souvenirs. The next news was that Rhapsody was indeed pregnant, with a full-term calf. It was a devastating blow to this population, particularly because we later found out the calf was female. The necropsy determined that the fetus had died first, and the resulting infection from a failure to expel the deceased fetus was the cause of Rhapsody's demise. I later asked Balcomb if emotion creeps in while doing a necropsy on a whale he's known for so long—in the case of Rhapsody, since birth—or if the scientist side takes over and he can objectively observe the grim scene. His response: "There were definitely moments of grief." As 2014 drew to a close, the whale-watch community had endured a period of twenty-eight months without a successful birth among the Southern Residents. L120 had been the first known calf in two years in August, but the calf only lived for seven weeks.

The data collection done by the Conservation Canine team involving Tucker, the scat-detecting black lab, and other dogs, has yielded additional insights into the issue of pregnancy among the Southern Resident killer whales. When fecal samples are collected in the field, it's often when a large group of whales are present, and since the research team doesn't follow right behind the whales, they don't know which whale their sample comes from. However, the researchers can screen DNA from the fecal sample against NOAA's genetic database for Southern Residents built from blubber biopsy samples. This allows them to determine, in most cases, which whale the sample is from. If a fecal sample is large enough to conduct hormonal analyses, researchers can screen for high levels of progesterone and testosterone as a sign of an advanced pregnancy. The really revealing information comes when looking at the corresponding nutritional hormone data from these same fecal samples. In the cases where the pregnancy hormone profile resulted in a successful birth, those females were in good nutritional condition at the end of the summer season of feeding in the Salish Sea. In the cases where no calf was observed and it is presumed a miscarriage happened, the females showed greater signs of nutritional stress. Of the thirty-five pregnancies detected in the Southern Residents over a seven-year period, only eleven of these pregnancies resulted in a documented birth.

This research is another example of the benefit of long-term monitoring projects among the Southern Resident community. These killer whales give us a unique opportunity among cetaceans to collect multiple samples from known individuals over time. This allows scientists to monitor the hormone levels (and thus nutritional and reproductive status) of particular whales. Unfortunately, no fecal sample had been collected from Rhapsody in the year preceding her death, but samples were collected from her in 2011 and 2012. She showed the pregnancy profile in 2011 but not in 2012, and she was never seen with a calf. This indicates that the fetus who died so close to full-term in early December 2014 was at least Rhapsody's second failed pregnancy. After the adult whale's death, a dark cloud hung over the whale-watch community. Another Southern Resident whale gone, and another new calf with her. Luckily, a break in the clouds was just a few weeks away.

On December 30, as J-Pod made its way north up Haro Strait, the Center for Whale Research documented a new calf with them—J50, later determined to be born to J16 Slick, a successful mother and now one of the oldest on record having given birth at the age of forty-two. This little calf, a female, was named Scarlet, in reference to the deep scratches and puncture wounds she showed shortly after birth. Researchers theorized these might have been indicators of a difficult entry into the world, where she was perhaps pulled out of the womb by another whale. Scarlet was just what we, and arguably the whales themselves, needed: a spunky survivor.

Throughout the following year, the whale community enjoyed what we all called a baby boom, with an incredible nine documented births in thirteen months. According to the Center for Whale Research, a spate of births like this hadn't been seen since the second year of their study, when nine whales were born in a twelve-month period in 1977. Finally, it seemed like things could be turning around for the Southern Residents. But why were there so many successful births all of a sudden, and were things really moving toward long-term improvement? With an eighteen-month gestation period, many of these new arrivals had been conceived during 2013, which interestingly was a year we didn't see much of the Southern Residents in the Salish Sea. While Fraser River Chinook runs were poor that year, returns to the Columbia–Snake River Basin were high, which is probably why the whales changed up their travel patterns. (Unfortunately, that was probably an artificially high return because that year was bolstered by a short-term hatchery project, not a trend of salmon recovery.) Presumably, the whales took advantage of the food abundance somewhere else and were able to eat enough to be able to successfully carry all these calves to term and nurse them after—a huge nutritional toll on the mothers.

L103 Lapis with her first-born calf, L123 Lazuli. Lazuli was part of the "baby boom" of 2015: of the nine calves born over a thirteen-month period, six of them were still alive two years later. All three of the L-Pod calves, including Lazuli, are males.

The best news was that three of these calves were born to young mothers in L-Pod, the pod in most dire need of successfully reproducing females. For successful recovery, the whales need not just to be born but to survive, grow up, and reproduce, and for many more calves to be born in the coming years. It's important to not take one good year as evidence that all is fine with the Southern Residents. Only about half of the breeding-age females in the population are producing offspring, and none of them at what is considered the expected rate of a new calf every three to five years. Another startling fact is that among the juveniles in K- and L-pods, there is a strong male bias. Of the thirteen calves of known gender born into L-Pod in the ten years from 2004 to 2014, just four of them were female, and only three of those are still alive today. Over the same time period, only one surviving female calf was born into K-Pod. The flurry of births in 2015–16 did little to change those numbers, as only two of the nine calves were female, both in J-Pod. The reason for the male-biased sex ratio among calves in K- and L-Pods may very well be tied in to their different toxin loads caused by metabolizing more toxins out of their blubber as they travel further than J-Pod or due to being exposed to different chemicals in their travels off the outer coast of California.

It took many decades for the Southern Residents to end up in their current predicament with their various threats of limited prey, toxins, and vessels. A full population recovery will be a long process, but it was a thrill to see nine babies, and it felt like proof from the whales that if they do get enough to eat, they will be able to successfully reproduce and recover. But we aren't there yet. Since the last calf of the baby boom was born in January 2016, there has not been another successful birth as of the end of 2018—another period of nearly three years without a new calf to celebrate. To add insult to injury, J50 Scarlet, the baby that kicked off the baby boom, gave so much hope to so many people, and was one of just two females born into the population in the last five years, died in September 2018.

IS THEIR POPULATION TOO SMALL?

With such a small population, the lack of genetic diversity and risk of inbreeding was identified by the US and Canadian governments as a potential confounding factor limiting the recovery of the Southern Residents. The number of breeding adults is already limited, and given the sometimes conservative culture of killer whales (e.g. only eating certain types of prey despite others being more widely available), there was concern that their breeding structure might further limit mating opportunities. We already know this to be the case on the ecotype and population level. Southern Residents are biologically capable of mating with transients or Northern Residents, but they don't because of their cultural rules. Does their mate choice within their population further restrict breeding opportunities?

To answer this question, we have to look further than observed sexual behavior. For instance, from behavioral observations of songbirds, it was believed about 90 percent of species were monogamous. With the advent of genetics, however, it turned out that just the opposite is true: about 90 percent of species are actually polygamous, they're just very good at hiding their polygamy from human observers. In killer whales, based on observed sexual behavior, they seem very promiscuous. Sexual interactions aren't restricted based on age, relationship between the individuals, or gender. One might infer from this that everyone breeds randomly with everyone else in resident killer whale society, but again, thanks to genetics, it turns out this isn't true. Mike Ford of NOAA looked for answers to these questions about Southern Residents' breeding structure starting in the mid-2000s. With long-term photo-ID and monitoring studies, the maternal relationships among the Southern Residents are known with certainty. Although relationships among individuals before the 1970s are inferred by association, every birth since 1976 has been carefully documented by the Center for Whale Research and thus we know the mothers of most of the current population. But getting a full picture of breeding structure would require knowing the paternity of the calves as well. NOAA scientists began collecting skin samples from the whales with biopsy darts to build a genetic database for the Southern Residents.

Researchers are able to determine paternity from the DNA extracted from these biopsy samples. Imagine a sample is collected from K42 Kelp, who was born in 2008. His cells contain paired strands of DNA (chromosomes), and he gets one chromosome from each of his parents. His mother is K14 Lea, so half of Kelp's chromosomes are from her. We know the killer whale gestation period is eighteen months, so we can go back to when Lea got pregnant and determine who all the potential fathers might be given which males were alive and of breeding age at that time. If the sample database is large enough, scientists can compare Kelp's DNA to that of all the potential fathers and see which one, when combined with Lea's DNA, is the best match to be his father. Following this process, it was determined that L41 Mega is most likely Kelp's father. Mega and Lea mating fits the pattern of the killer whale breeding structure seen in the Northern Residents, where genetics show the whales mate outside of their own pod. The Southern Resident population is much smaller, however, and as more fathers were determined by this method, it was found that a pattern of outbreeding to other pods doesn't hold up. Successful adult breeding males will mate with members of all three pods—Js, Ks, and Ls—including their own.

Another interesting (though somewhat discouraging) pattern emerged as more paternities were revealed: Not all the males that are capable of breeding are successfully fathering offspring. If potential fathers are considered to be males age fifteen or older (the same age females tend to reach sexual maturity), there are often twenty or more potential fathers among the population, but the vast majority of calves in recent years have been traced back to just two males: J1 Ruffles and L41 Mega. They have each fathered more than fifteen offspring, and more than 50 percent of the current population is descended either directly or indirectly from Ruffles or Mega. Some facet of Southern Resident breeding structure is therefore limiting mate choice. Females are not rearing offspring randomly fathered by any male in the population. Given that Ruffles and Mega are both the oldest and the largest males within the population over the study period, it's likely that either these males are dominating others for access to mates or that females are choosing them to be fathers. Since there haven't been any observed signs of male-male aggression, and we know that killer whale societies are matrilineal and likely matriarchal, female mate choice seems most likely to be driving this phenomenon. For some reason, older (or perhaps larger) males appear to be the preferred mates. One reason for this preferential mating with older males might actually be to avoid inbreeding. Since neither males nor females disperse from their natal pods—an anomaly in the animal kingdom—there's risk of close relatives mating with one another. However, if a mother's son won't start breeding until he's about thirty, she'll likely be postreproductive by then. Similarly, if a father is over thirty when he has a daughter, he'll likely be deceased by the time she reaches breeding age. The majority of known matings have occurred between two whales who, given their ages, could not possibly be parent-offspring.

As a result of genetic paternity tests over the past decade, Mike Ford and colleagues have discovered four known cases of inbreeding among close relatives of Southern Residents. Given their small population, there are likely many more cases of inbreeding among more distant relatives, but so far, individuals don't seem to be suffering from the consequences of this inbreeding, which is actually similar to the rate of inbreeding seen in other species. Inbreeding is a concern not only because it lowers overall genetic diversity, but because of inbreeding depression. Deleterious recessive alleles are harmful versions of a gene that won't get expressed if you only have one copy of them. With inbreeding, however, it's more likely that an individual will end up with two copies of these genes from their closely related parents and therefore experience an overall decrease in fitness compared with other cross-bred individuals. So despite the evidence of inbreeding among closely related individuals in Southern Residents, it does not seem to be occurring at an alarming level. What all this suggests, according to Mike Ford, is that the benefits for resident killer whales of staying with their mother's group are likely outweighing the costs of the risk of inbreeding.

Known Cases of Close Inbreeding in Southern Residents

Mother + Father = Offspring	Relationship between Parents
J16 Slick + J26 Mike = J42 Echo	Mother and son
J28 Polaris + J1 Ruffles = J46 Star	Father and daughter
K13 Skagit + L41 Mega = K34 Cali	Half-siblings (shared father)
K22 Sekiu + L41 Mega = K33 Tika	Half-siblings (shared father)

Source: Michael J. Ford, K. M. Parsons, E. J. Ward, J. A. Hempelmann, C. K. Emmons, M. Bradley Hanson, K. C. Balcomb, and L. K. Park, "Inbreeding in an Endangered Killer Whale Population." Animal Conservation (2018). https://doi.org/10.1111/acv.12413.

Killer whales as a species have overall comparatively low genetic diversity compared with other species. While the Southern Residents are at an extreme (with fewer than 80 animals in their current population), in general, killer whales have fairly small population sizes throughout the world. The global population of killer whales is high, but because of their segregated nature, most breeding populations are likely on the order of only a few hundred animals. Since small populations are at a greater risk for inbreeding and low genetic diversity, I always figured that killer whales must have some as-yet-unidentified mechanism for biologically coping with this. As of 2017, one such mechanism may have been identified in the preliminary research findings of Lance Barrett-Lennard and his colleagues. Many genes lead to traits that directly influence how well an individual can survive and breed, and if a small population starts to lose genetic diversity in genes that help fight diseases, for instance, it will be more vulnerable as a whole to those diseases. How do killer whales, given their small populations, deal with this? Individual orcas may carry *more* than two copies of each gene. Looking at a known immune system gene in Southern Residents, Barrett-Lennard found individuals carrying up to twelve copies of the gene! The actual genetic diversity of killer whale populations might therefore be much higher than we would traditionally think.

Barrett-Lennard's cutting-edge genetic research is likely to influence our thoughts on the recovery possibilities of the Southern Residents and other small orca populations. Despite having only a few dozen active breeders among the Southern Resident community, Barrett-Lennard does not believe they are doomed based on genetics.

LEARN MORE

For more on the habitat shifts of the Southern Residents and their increasing absence from the Salish Sea, see data published on the website for the Center for Whale Research and the 2018 paper published by the author and colleagues from the Orca Behavior Institute. See the publications by Sam Wasser and Jessica Lundin for more on the Southern Residents and pregnancy. Mike Ford and Lance Barrett-Lennard have published fascinating studies on killer whale genetics.

CHAPTER 12

Saving an Endangered Population

WHEN THE SOUTHERN RESIDENTS WERE listed under the Endangered Species Act in 2005 criteria had to be put in place by NOAA for what would allow their de-listing. These benchmarks include having a population structure similar to the Northern Residents in terms of postreproductive, breeding age, and juvenile whales of both genders and having experienced a growth rate of 2.3 percent for a period of twenty-eight years—deemed to be equivalent to two killer whale generations. For a population of about eighty animals, this is equivalent to growth of the population to about 150 whales over those 28 years. Unfortunately, over the first ten years of their listing, the Southern Resident population continued to decline.

THE FIRST TEN YEARS

In 2014, NOAA released a ten-year summary of research and conservation of the Southern Residents, in which they detailed their work to date, including expenditures. Of the $11.9 million spent on Southern Resident science and research between

2003 and 2012, 33 percent of the funds went to studying seasonal distribution (primarily via satellite tagging) and 18 percent to vessel noise and interactions—the top two spending categories. Just 9 percent of the funds went to prey studies. Of the $3.8 million spent on management and conservation actions over the same period, most of these funds went to contract support (28 percent) and education/outreach (21 percent). Vessel interaction and noise received 20 percent of these funds, while prey issues received just 7 percent.

NOAA's described goals in the report for the next ten years across all categories are to continue studying and monitoring the relationships between the orcas and salmon, toxin levels, and vessel effects. While they stated their intent to continue exploring potential additional conservation actions, the only one they specifically identified is possibly creating a protected area, presumably from boats.

A spectacular breach by J19 Shachi. Since the death of J2 Granny, Shachi is often in the lead when J-Pod travels the Salish Sea. With the Southern Resident population decline, Shachi faces a tough task of leading her family to abundant food.

It is my opinion that NOAA has done an excellent job in conducting solid science to help fill the identified data gaps, but I see substandard performance in implementing effective management actions that will help with recovery of the Southern Resident population.

In 2015 the Southern Residents were identified in a biennial report from NOAA to the US Congress as one of eight "Species in the Spotlight." The report named species at greatest risk of extinction if immediate actions aren't taken. Part of a result of this listing is that NOAA had to develop an additional five-year action plan on top of the existing recovery plan for the species. The key actions they identified in this 2016–20 action report are: (1) "protect killer whales from harmful vessel impacts through enforcement, education, and evaluation"; (2) "target recovery of critical prey" by continuing to study seasonal diet, developing a model of how killer whale population dynamics respond to other predators' impacts on Chinook, and evaluating the importance of hatchery stocks in the Southern Resident diet; (3) "protect important habitat areas from anthropogenic threats" by continuing to consider the expansion of critical habitat; (4) "improve our knowledge of Southern Resident killer whale health to advance recovery" by continuing photo identification, photogrammetry, and fecal sampling; and (5) "raise awareness about the recovery needs of Southern Resident killer whales and inspire stewardship through outreach and education."

Where on this list is the dedicated action that makes a real effort to provide more salmon for these whales? Several longtime whale scientists and observers offered their reactions. Ken Balcomb detailed his thoughts in an open letter in which he stated that NOAA "has been very busy cranking out 'paper salmon' and 'paper whales' with state of the art math-magic that may demonstrate all is well and on course with salmon and Southern Resident killer whale recovery efforts; but, critique of the data details and machinations aside, it is obvious that all is not well with Southern Resident killer whale recovery."

Howard Garrett of Orca Network said: "I see a lot of ways of looking at and studying the problems, forming partnerships, and doing some education, but other than enforcing whale-watch regulations, I don't see much action on the ground. If only NMFS [the National Marine Fisheries Service] could weigh in on the controversies about dams on the Snake River, for example, or enforce quieter engines on cargo shippers, or help make the case for protecting whales from naval training exercises. I think it's within their legal purview to apply their statutory authority to help resolve some of these complex issues in favor of the life of the biosphere, without having to be sued to do it."

Brian Goodremont, a whale-watch tour operator based on San Juan Island, had this to say: "I think it's just sad that an endangered species like the Southern Residents isn't enough to affect salmon policy. I've had several managers basically say that 'Southern Resident killer whale endangered status will not affect salmon recovery,' meaning they don't have the power to effect change with salmon policy. So what's the point of the ESA mandate then? We know more salmon within the critical habitat of Southern Residents will help their population grow. I also look at the amount of government waste and think, *Why not put a moratorium on catching certain salmon stocks for five years and see what happens?* Reimburse fishermen for their lost catch on an escalating curve so they receive 100 percent of the mean average reported catch from the last five years. Reimburse the canneries and the supply line. See if we can recover the salmon for the salmon's sake, and recover Southern Resident killer whales." He added: "It's discouraging to see a lack of effort from NOAA to effect change with salmon catch policy and restoration efforts to benefit Southern Residents. We need long-term restoration of habitat but short term we just need more fish for the whales until some of our ten-, twenty-, thirty-, and forty-year restoration plans start taking hold."

When the Southern Residents were listed as endangered in the US, there were 87 of them. Thirteen years later, at the end of 2018, there were 74 Southern Resident killer whales. That's a net loss to the population of one whale per year. No matter the metric, it is clear we have not yet done nearly enough.

THE END OF A DAM ERA

At the ten-year anniversary of the ESA listing of the Southern Residents in 2015, their population had declined a further 10 percent. Of the three risk factors identified— prey availability, toxins, and vessel effects—NOAA took the most action on vessel effects, by instituting new vessel regulations and considering a no-go zone. Although NOAA sponsored important research to fill the data gaps, there were no actions on the horizon to use this data to get the whales more food. In response, dedicated whale advocates banned together to figure out what they could do to generate more Chinook salmon for the whales, rather than relying solely on government action.

Out of this sentiment emerged a series of CALF (Community Action – Look Forward) workshops, which included breakout groups for brainstorming ways to tackle increasing action on the salmon issue on several fronts: education, research, and action. The education group devised efforts for spreading the word about the orca-salmon connection to get more people involved, such as linking whale adoptions to local stream adoptions and developing a handout summarizing the issue. The research group discussed where the unknowns were and who was working on them, such as the Salish Sea Marine Survival Project, which looks at what is affecting juvenile salmon survivorship in the region. The action group focused on direct things that could be done to advocate for more salmon for the whales, such as forming a political lobbying group and communicating the economic importance of the orcas to the policy makers.

One of the major frustrations expressed at the first (US-based) CALF workshop was that the main summer source of food for the Southern Residents is Canada's Fraser River, over which Americans have little influence. While the goal was for the workshops to expand to be transboundary, an immediate focus point was thus what could be done in the United States. A major salmon source within the United States is the Columbia–Snake River Basin, an area US government officials help manage, and here we found an issue we could speak up for on behalf

of the orcas: dam removal. In particular, the four Lower
Snake River dams are widely recognized for dramatically and negatively impacting salmon returns. Originally
built to provide hydropower and open up navigation to
Lewiston, Idaho, the dams, while still important to some
locals, provide continually declining benefits and increasing costs. In terms of energy, they produce most of their
power in late spring and early summer, when energy supply is high and demand is low. Overall, the dams produce
at about a third of their nameplate capacity, representing
less than 4 percent of the power produced in the Pacific
Northwest. Wind, solar, biomass, and conservation and
efficiency upgrades have already replaced hydropower
from the Lower Snake River dams several times over with
affordable and renewable energy—a trend that promises
to continue. Between 1995 and 2015, freight transport on
the Snake River has declined more than 70 percent, much
of it replaced by rail. The Army Corps of Engineers now
classifies the Snake River as a waterway of "negligible use."
With usage declining, infrastructures aging, and the dams
slated for extensive (and expensive) maintenance billed to
taxpayers in the coming years, it makes sense on all fronts
for these dams to go.

Here was an action whale advocates could focus on that
would have real and immediate impact for the Southern
Residents; if we could get these dams breached, there's
potential to restore millions of fish into the Pacific Northwest
ecosystem. According to Save Our Wild Salmon, removing
the four Lower Snake River dams would open up more than
fifteen million acres of prime salmon habitat. Since several
Snake River salmon stocks are endangered, with the first

being listed in 1991, NOAA has been required to develop recovery plans for these fish. In the nearly three decades since then, none of the recovery plans have made it out of draft form, as they keep being rejected by the courts as illegal and failing to meet the required mandates. Meanwhile, the fish, and the whales, continue to suffer. US District Judge James Redden, who presided over ten years of the litigation regarding the Columbia–Snake River salmon recovery plans, spoke bluntly to the media after his retirement. "I think we need to take down those dams," he said. Nearly all the technical reports published within a decade or so of the endangered listing of the salmon, which looked at the benefits of various recovery options including dam breaching, concluded that removing the four Lower Snake River dams would be the most certain and effective route to salmon recovery.

The environmental benefits of dam removal used to be largely unknown, but many smaller dams have been removed, and the resulting ecosystem recovery has occurred faster and more extensively than expected. The largest dam removal project in the United States to date occurred right in the core Southern Resident habitat when in 2014 the last piece of concrete was removed from the Elwha and Glines Canyon dams on the Elwha River, opening up pristine salmon spawning habitat in Washington's Olympic National Park. After one hundred years of being blocked, salmon returned to their historic spawning grounds faster than anyone could have predicted. Just two years after dam removal was complete, Chinook salmon were spawning in the river above the former Glines Canyon dam site. While the twentieth century boasted the era of dam building, it's beginning to look like the twenty-first century will be a period of dam deconstruction But will it happen fast enough to help the Southern Residents?

Out of the CALF workshops formed a one-issue group called the Southern Resident Killer Whale Chinook Salmon Initiative (of which I am a part), whose mission is to advocate for more salmon for the orcas with a strong focus on lobbying for Snake River dam removal. As the group met with politicians in both Washington State and Washington, DC, they basically said that they agreed with our concerns, but it was unlikely there was anything they could do. In 2014, we turned our hope to convincing President Obama to sign an executive order to breach the dams, and what followed was a story of behind-the-scenes political drama worthy of its own book. Unfortunately, despite continued assurances that key people were on board with the dam breaching, the advocates' efforts did not come together in the midnight hour of the Obama administration to pull off the executive order.

Although many scientists, engineers, economists, and technical experts agree that these four dams in southeast Washington (the Lower Granite, Little Goose, Lower Monumental, and Ice Harbor dams) need to go, the issue has been bogged down because of politics. A small number of dam proponents, including the Bonneville Power Administration (BPA) that sells the region's hydropower, have worked to keep any discussion of dam removal off the table. In 2008, for example, the BPA, US Army Corps of Engineers, and Bureau of Reclamation signed the Columbia Basin Fish Accords with regional tribes, which funded habitat restoration efforts in exchange for the tribes agreeing not to advocate for dam removal or increased spill over the dams for a period of ten years. In what is historically a pro-dam state, many legislators and members of the public are concerned that if some dams start to be removed, environmentalists will want to take all the dams down. With congressional gridlock, the chance of any legislative action being taken to breach the dams is slim to none. Multiple interest groups including farmers, tribes, and fishermen who were long at odds with one another came together and signed an agreement in 2010 that included the removal of four dams on the Klamath River in California, but progress has been stalled for years by Congress's failure to approve funding for the project. The situation on the Snake River is just as complicated. It's more than just the Cascade Mountains that divide eastern and western Washington. The state is polarized on the dam issue, with the eastern Washington farmers and irrigators staunchly in favor of the status quo and the urban environmentalists in the western part of the state putting a higher value on salmon recovery as a priority. Until this divide is breached in the form of finding mitigation solutions to make all parties whole moving forward, dam breaching remains a pipe dream.

One result of this increased advocacy for dam breaching on behalf of the whales was a white paper issued by NOAA in 2016. Despite saying in their 2008 Southern Resident recovery plan that "perhaps the single greatest change in food availability for resident killer whales since the late 1800s has been the decline of salmon from the Columbia River Basin," their Lower Snake River dam white paper backpedaled. "No one salmon recovery action on a single river, such as breaching dams on the Snake, would itself bring about the recovery of the Southern Resident killer whales . . . however, long-term recovery of West Coast salmon and their habitat collectively, including Puget Sound, will *likely* [emphasis added] be an important contributor in the recovery of the whales." One month after this position paper was released, Ken Balcomb of the Center for Whale Research and others sent a letter to NOAA asking for a retraction, since they found it unsupported by the best available science. "According to Steven Hawley's book Recovering a Lost River, between 2001 and 2011, NOAA's Northwest Fisheries Science Center, which is responsible for both endangered salmon in the Columbia-Snake River Basin and the endangered Southern Resident killer whales, received more than three-quarters of its budget from the Bonneville Power Administration and the Army Corps of Engineers. This ethically compromising fact was not disclosed in their white paper."

As of 2018, the Snake River salmon recovery plans are back in court, and dam breaching continues to be discussed with no concrete actions being taken toward actual dam removal. A court-mandated Environmental Impact Statement and biological opinion on the Columbia hydropower system, including Snake River dam removal, is underway, with a decision slated to come in 2020. While this sounds promising, many are criticizing it as an unncessary delay and a replication of a previous effort that concluded with a 2002 study that stated Snake River dam breaching was the best option for salmon recovery. Public momentum in favor of dam breaching is continuing to build; a petition started by the Southern Resident Killer

Whale Chinook Salmon Initiative urging politicians to support Snake River dam breaching is one of the top ten all time petitions on change.org, surpassing 700,000 signatures by the end of 2018. Meanwhile, the orcas continue to search for fish.

SETTING A NEW COURSE FOR RECOVERY

Breaching the Lower Snake River dams has been a strong focus of advocacy efforts on behalf of the Southern Residents, but there are other options that could make a real difference toward their recovery. The continued decline of the Southern Residents despite their endangered listing has slowly garnered more attention. By the end of 2017, the population numbered just seventy-six whales—a thirty-year low. This shocking fact led to a series of workshops to address emergency actions that could be taken to reverse this trend, including one by the federal government of Canada in Vancouver and by San Juan County in Friday Harbor, Washington. New actions are required at all levels of government—from local to federal.

One major concern is that the Southern Residents are not given a seat at the salmon management table, so to speak. When fisheries managers are dividing up "take" for salmon, if the energetic needs of the whales are considered alongside those of commercial, recreational, and tribal fishermen, it might result in the orcas receiving a larger piece of the pie. Fisheries management is incredibly complex, but as of 2018 no governing agencies consider the Southern Residents and their salmon needs when allocating salmon harvest quotas. This could happen at the state or provincial level, the federal level, or even internationally via the Pacific Salmon Commission, which coordinates between the United States and Canada. A precedent for giving predators a quota in fisheries management was set in early 2017 when the National Marine Fisheries Service requested that the Alaska Department of Fish and Game consider the dietary needs of endangered Cook Inlet belugas in their salmon management plans.

It's not just the endangered orcas that need to be considered, either. For several decades those modeling fisheries and trying to set quotas mainly considered oceanic conditions and human take. Other species such as seals and sea lions that historically ate large quantities of salmon weren't an issue when salmon modeling efforts began because their populations were so low. New studies have shown, however, that these recently recovered marine mammal populations play a bigger role in salmon numbers than managers realized. Over the past forty years, Chinook salmon recovery efforts have increased and human take via fisheries has declined, but instead of these facts translating into larger spawning populations of salmon or more killer whales, the result may in fact be the large seal and sea lion populations we now see along the West Coast. The greatest number of Chinook salmon are consumed not by humans or orcas, but by harbor seals, which eat juvenile salmon before they become large enough to be of interest to fishermen or killer whales. Going forward, we must consider all salmonid predators when modeling and managing salmon populations. The fact that seals and sea lions are eating so many fish has brought back old arguments for culling these species.

Culling was tried before, which is why seal and sea lion populations were so low in the 1970s, but it had no documented effect of increasing fish numbers for fishermen. Biologists caution us not to make the same mistakes again: Given a little time, the growing marine mammal–eating transient killer whale population will balance out the booming seal and sea lion numbers. Indeed, over the last 15 years, transient killer whale presence has continued to increase in the Salish Sea, and harbor seals numbers have been declining in both British Columbia and Washington.

A harbor seal eats a salmon.

With fisheries management and salmon recovery issues so convoluted, one novel approach has been gaining traction in recent years: Instead of supplementing fish populations via hatcheries for humans, what if we do so for killer whales? There aren't enough wild Chinook salmon to support commercial, recreational, and tribal fisheries, let alone to support the dozens of other species that also feed on Chinook. Although hatcheries in native salmon rivers can be problematic to wild fish by outcompeting them or reducing genetic diversity, they can be sourced from native stream stocks and may be necessary in the short term to help boost the number of salmon in that habitat for everyone.

In addition to boosting production at existing hatcheries, another potential solution is to locate hatcheries in areas where Chinook don't traditionally spawn. The regional organization Long Live the Kings, focused on Chinook salmon recovery, is one of the major partners in the Salish Sea Marine Survival Project. The project studies issues detrimental to juvenile salmon, and started an experimental hatchery in 1978. Located on Orcas Island, where there are no salmon rivers, the Glenwood Springs Hatchery releases up to 750,000 juvenile Chinook a year into the ecosystem, with one of their stated goals to help feed the Southern Resident killer whales. What if more projects like this happened on a larger scale? In 2017 the Southern Vancouver Island Anglers Coalition, with support from Fisheries and Oceans Canada, took another step. Fishermen and whale-watchers came together to fund the release of 220,000 juvenile Chinook from Sooke Harbor, with plans to scale up to a million fish a year by 2020. The anglers' coalition president, Christopher Bos, said: "The community is coming together to do something that the government can't afford—to help an endangered species find food certainty—and to benefit a community that relies on the business of recreational fishing." Such citizen-led initiatives are key if the Southern Residents are going to have a chance.

A few other pioneering efforts that could be precursors to major changes have also begun throughout the region. The Vancouver-Fraser Port Authority in British Columbia has an Enhancing Cetacean Habitat and Observation (ECHO) Program to mitigate the impacts of shipping activities on marine mammals. In 2017 they sponsored a voluntary vessel slowdown in Haro Strait primarily to assess the noise reduction impacts of reducing speed of shipping traffic in the region to a target of 11 knots (a knot is equal to one nautical mile per hour). More than 60 percent of commercial vessels complied with the slowdown during the two-month trial. Using the hydrophone off Lime Kiln Lighthouse to measure ambient noise levels, scientists reported a 24 percent reduction in sound intensity during the trial period. When looking specifically at times within the trial where small boats weren't present and weather conditions weren't likely contributors to increased ambient noise, the overall sound intensity drops from shipping traffic alone reached a 44 percent reduction. The ECHO program is continuing to monitor and experiment with the most effective ways to reduce underwater noise disturbance from shipping traffic in the Salish Sea.

Also in 2017, a new population viability analysis of the Southern Residents was released to assess what it would take to help the population achieve growth levels in line with recovery goals. The study, authored by Robert Lacy and colleagues, found that by addressing prey management alone, regional Chinook salmon levels would have to increase to numbers not seen in the past fifty years. However, by addressing multiple factors, the goals become more attainable. The study suggests that a 50 percent reduction in acoustic disturbance (for example, by slowing down shipping traffic through efforts like the ECHO program) combined with just a 15 percent increase in regional salmon numbers could also help the whales attain healthy population growth. While these numbers can't be taken as hard facts as they are theoretical models based on assumptions with varying degrees of certainty, studies like these lay the groundwork for attainable action plans; now we just need to see the action.

The Government of Canada showed a renewed commitment to action in 2018. The Department of Fisheries and Oceans (DFO) announced fishing closures in three areas known to be foraging grounds for the Southern Residents as part of a plan to reduce the British Columbia Chinook salmon harvest by at least 25 percent for the year. While the plan sounded promising on the surface, it was unsurprisingly controversial, particularly among the sport-fishing community in the Salish Sea. The plan didn't include any compensation for local fishermen, harming local businesses, and some feel the plan ignored the major recovery efforts done by fishermen to smaller regional rivers. In fact, in an ironic twist, implementing fishing closures actually harmed salmon habitat recovery efforts: as a result of the closure zones, several fishing derbies were canceled, and in turn, at least one salmon habitat recovery project was delayed without the much-needed funds from entry fees for the derby. One Vancouver Island fisherman said, "Habitat destruction, baitfish reduction, and noise pollution have all had a greater impact—and will continue to have—than recreational fishing that removes less than 1 percent of [the Southern Residents'] yearly diet. . . . The recreational contribution to habitat and stock rebuilding in Georgia Strait is in the millions of dollars. . . . We are feeling extremely insulted and marginalized."

Another action taken by DFO in early 2018 found broader support: the designation of more than $9.5 million toward eight Chinook salmon habitat restoration projects across British Columbia, including juvenile salmon habitat at the mouth of the Fraser River and key spawning grounds on a major tributary to the Fraser, the Thompson River. After another difficult summer for the Southern Residents, in the fall DFO announced additional actions on behalf of the whales, including further reductions to Chinook salmon fishing, an increased investment in Chinook hatchery production, and a plan to extend critical habitat for the whales beyond the inland waters of the province to include areas on the outer coast.

On the other side of the border, Governor Jay Inslee of Washington State signed an executive order in early 2018 to jumpstart immediate and long-term recovery efforts for both Chinook salmon and the Southern Resident killer whales. One of the directives was to boost Puget Sound hatchery production of Chinook as an immediate stopgap. The other main thrusts of the executive order were additional funding for enforcement of vessel regulations and the establishment of a task force made up of members of multiple interest groups to make recommendations to the governor for additional actions that could be taken at the state level or by partnering across multiple states and provinces. This was a positive sign; if the federal government won't make the difficult decisions to take dramatic action, perhaps state governments will. In the inaugural meeting of the state task force, Inslee made the promising statement of telling the members that no possible actions should be considered off the table. Year 1 of the task force meetings had mixed results. On one hand, working groups made science-based recommendations for bold changes in the state regarding Chinook salmon abundance, contaminant levels, and vessel impacts; these recommendations were further refined by the task force and via public feedback into a solid package to present to the governor. On the other hand, politics and special interests prevailed. Some task force members openly stated they were there to protect the interests of their stakeholders, despite the mandate to put the whales first. Others made backroom deals for political interests, wheeling and dealing one action item for another, again with little regard for the science or the whales. One example was an 11th hour proposal for a moratorium on whale-watching of the Southern Residents, a recommendation the task force voted to proceed with in the final meeting of the year after only a few minutes of deliberation, with no scientific or public input, and without any discussion of the implementation details. While on paper this looked like a bold recommendation by the task force, in reality it could do more harm to the whales than good. Without professional whale-watch boats on the water, research and enforcement vessels won't having sightings input on where to find the whales and private boaters won't have a visual on whale locations either, increasing the likelihood of high risk infractions of vessel regulations (for instance, boats driving right over the whales because they don't know they are present). Additionally, the Pacific Whale Watch Association plays an important role in alerting the military to killer whale presence in active testing zones; their reports put a halt to potentially dangerous exercises that could harm the whales when they are present. The PWWA filed a minority report signed by more than 10 regional killer whale scientists and 8 regional killer whale organizations stating their opposition to the idea of a moratorium as providing benefits to the whales, providing a striking example of how the task force failed to listen to the most knowledgeable experts in the region with regards to what is best for the Southern Residents, instead opting for political victories.

At the time of publication, Governor Inslee received the Year 1 task force recommendations and used them to craft a 1.1 billion dollar proposed budget for orca recovery that the local media heralded as "billion dollar bold". While this indeed is an impressive figure, the details were less so. More than a quarter of the proposed budget ($296 million) was slated to go towards culvert replacement in Puget Sound, an important action to improve salmon access to habitat, but an action already mandated earlier in the year by a Supreme Court decision ruling in favor of 21 tribes that had been in a long-fought battle with the state over this issue. Another large line item was the $117 million out of the transportation budget to electrify ferries, converting two existing ferries and building two more in the name of quieting waters. Given the number of large ships operating in the region this will be of minimal impact to the whales, and is another case of double-counting funds as benefitting orcas when it's really a part of Inslee's clean-energy plan to combat his key issue of climate change. Even the largest line item on the proposed budget, an impressive

$376 million for salmon habitat restoration projects, is less impressive in the details. While some programs will get a boost, this falls far short of the task force recommendation to fully fund proposed habitat restoration projects in the state, the most widely supported action item on their list of more than thirty recommendations. Many programs will see partial funding similar to the previous biennium, meaning many vetted, shovel-ready projects will continue to be on hold. Inslee also made the disappointing choice to move forward with the moratorium on viewing Southern Resident killer whales.

Throughout the process, the loudest public comments came with regards to breaching the Lower Snake River dams, and this public pressure ensured that the dams became part of the conversation after not being on the table in the earlier meetings. The task force recommended the convening of a stakeholder panel to discuss the costs and benefits of potential dam removal and develop mitigation options in the event of dam breaching, the idea being to supplement the ongoing federal review process of salmon in the Columbia Basin which will include another look at breaching the Lower Snake River dams. While this is an important step in bridging the divide between eastern and western Washington, it remains to be seen if a concrete mitigation plan will indeed be developed, or if it will just be another way for both sides to vent their concerns with no concrete progress towards much-needed solutions.

There will always be barriers to major changes, but the whales need more dramatic actions taken. Big efforts take time, but the clock is ticking. More research and education efforts will only take us so far, and while there are promising conversations happening to take renewed action, we have yet to see how it will play out. It's been more than ten years since the whales were listed as endangered, and what progress have we made on ensuring these whales have abundant prey and clean, quiet waters? How long do we have until it's too late?

A tight group of Southern Residents surfaces in unison in the calm waters of the Salish Sea.

LEARN MORE

For more on federal recovery efforts to date, see the National Marine Fisheries Service's *Southern Resident Killer Whale Recovery Plan* and *Southern Resident Killer Whales: Ten Years of Research and Conservation*; and Fisheries and Oceans Canada's *Recovery Strategy for Northern and Southern Resident Killer Whales*. To learn more about dam breaching, see the documentary films *DamNation* and *Return of the River*. The issue of salmon recovery and the Snake River dams is covered in detail in *Recovering a Lost River* by Steven Hawley. The coalition DamSense (www.damsense.org) advocates for breaching of the four lower Snake River dams and provides an extensive source of information on the issue on their website. Look into the Port of Vancouver's ECHO Program to learn more about vessel speed trials and other efforts by the port to improve Southern Resident killer whale habitat. Details on British Columbia's renewed commitment to salmon habitat restoration can be found on the project site for DFO's Coastal Restoration Fund. The Governor's Orca Task Force and many other programs are still actively unfolding. Southern Resident killer whale recovery remains a hot topic and is regularly in the news. I have been and will continue to post the latest updates and calls to action on the Orca Behavior Institute Facebook page.

EPILOGUE

A Future for Killer Whales in the Northwest

I RECENTLY REREAD THE LAST chapter in Erich Hoyt's book
Orca: The Whale Called Killer, originally published in 1981.
Titled "A Future for Killer Whales in the Northwest," the
chapter considers the challenges facing orcas. I reflected on
how much attitudes toward orcas changed during the 1970s,
the subject of Hoyt's book. The main issues then were the
ongoing live captures, the opening up of the last true wilder-
ness on Vancouver Island to commercial logging, the push
to create a sanctuary for the killer whales, and the fact that
as the region's population grows and the whales' environment becomes more accessible, there would be increasing
contact between killer whales and people. Hoyt expressed
both worry and hope about this last concern; more people
around the whales meant more pollution, habitat degrada-
tion, and disturbance, but it also meant more interest in
and love for the whales, perhaps leading to protection of
their habitat and prey. It's interesting to reflect on where the
whale community and the orcas themselves are now, closing in on forty years after the publication of Hoyt's book.

Although there were still permits out there to capture wild
whales in British Columbia, the final captures in the Pacific

Northwest had in fact taken place by the 1980s. Other orca populations, particularly those in Iceland and Russia, were targeted next for the live capture industry, but the removals from North American populations were over. This isn't to say the effects of the capture era had ended; the reality of losing so many young whales from the same age class still affects the ability of today's Southern Residents to recover. Much of the habitat Hoyt explored in the 1970s is no longer pristine, although some of the battles to create nature reserves and sanctuaries were successful. Since that era, more people have experienced wild killer whales, the results of which has indeed been mixed. The wild whales of the Pacific Northwest have more advocates than they have ever had, but simultaneously they face more challenges than ever. The threats of prey declines, toxins, and vessel disturbances are real, to say nothing of climate change and ocean acidification.

So what is the future for killer whales in the Northwest? Although they face their own set of threats, the future for the Northern Resident and transient orca populations seems secure. The Northern Resident killer whale population has doubled since the 1970s; the transient killer whale population has doubled since 1990. These two thriving populations, which experience many of the same risk factors as the Southern Residents, are indicators that it is possible for people and killer whales to coexist. For the Southern Residents, however, their future is much less certain. "I don't think their chances are good," said Dave Ellifrit of the Center for Whale Research. "We're going to hover around sixty animals for a while. It's hard to remain positive when you look at the family histories and project the best-case scenarios [for more calves]—and we haven't had any best-case scenarios happen. . . . Toxins aren't going to get better anytime soon. . . . I don't think anything is going to change unless there's a huge influx of new fish."

Howard Garrett of Orca Network agrees. "Right now, things look fairly dismal," he said. "Eighty is a precarious number. They have fallen so fast from their modest incline. Political decisions on Chinook for the next several decades are going to be key. If the right people do the right things, the outlook for Southern Residents could change quickly." Everyone I talked to while writing this book felt the same way: We're at the tipping point for the Southern Residents. We're close to the point where we may lose them forever, but most everyone agrees there is still a chance to save them. "They do have a prospect for recovery, but it's going to be challenging," said NOAA scientist Brad Hanson. "They have a chance to be fine, but we're not too many stochastic events from this thing going the other way."

Most observers find hope in the resiliency of nature. When decisions are made to protect nature and help wildlife recover, results often occur faster than people think possible. Many advocates point to the success of the Elwha River dam removals as a hopeful symbol, where salmon returned almost instantaneously to spawning grounds that had been blocked for a century. "We're living in an amazing part of the world with a phenomenal creature as our neighbor. The world is changing before our eyes and most of us aren't even aware of it," said Ken Balcomb of the Center for Whale Research. "Killer whales are the best indicators of a healthy ecosystem, and if we ignore them, we're cutting our own survival. We're not that different from them biologically. But there is still a huge potential for things to recover."

The looming question is: will humans take enough action quickly enough to give the orcas a fair chance? Twenty years after he wrote the conclusion for his book, Hoyt wrote an epilogue for the latest edition. When he originally wrote the book, he had no way of knowing that by the year 2000 he would still be able to visit the very same whales in Johnstone Strait. I know that further south, I will have the opportunity in twenty years to visit some of the same whales in Haro Strait, but their future is murky. How many of them will still be here, and how many new orcas will have joined their ranks?

It seems that as a species we wait until our hand is forced to take necessary action. We're not very good at taking the seventh-generation mentality, the Iroquois philosophy that dictates decisions be made with sustainability and seven generations into the future in mind. Society is driven by instant gratification, the bottom line, and an act-now-ask-questions-later mind-set. If we continue waiting until the last moment to try and save these endangered species, we are gambling with their next seven generations as well as our own. Yet I see evidence that the culture could be changing. My generation is the first to be raised with environmental stewardship as part of our education, and we have yet to see the full impacts of this. The previous generation of whale scientists and lovers were part of a transformation: the image of killer whales changed from ferocious, frightening beasts to animals of charisma, worthy of our interest, study, and respect. I look around at my peers, my fellow whale scientists and whale lovers, and believe we are capable of helping the next transformation take place: seeing whales as animals of another culture, worthy in their own right, with rights of their own, that we should do everything in our power to protect.

J17 Princess Angeline and her daughter, J53 Kiki. The recovery of the Southern Resident population hinges on productive females like Princess Angeline raising daughters that become successful mothers.

September of 2018 was the only month of the year that felt somewhat normal when it came to the whales. With the drama of the attempted intervention to save J50 Scarlet, the 17-day vigil of grieving mother J35 Tahlequah, and the constantly moving targets of new proposed actions by the federal government of Canada and Washington Governor Jay Inslee's task force, there was more attention on the Southern Residents in the media and on social media than ever before. Meanwhile, in what appears to be the "new normal", the whales themselves were scarce throughout the season. For the first time on record the Southern Residents weren't present in inland waters at all during the month of May, and were in the Salish Sea less than half as much as historically during June, July, and August. But by September, members of all three pods were making their usual route to the Fraser River on a daily basis, and on one evening my husband and I caught them as they made their way down San Juan Channel at sunset.

All of J-Pod, all of K-Pod, and most of L-Pod were present: 59 whales in all, and the closest we had come to having a true superpod in the waters around the San Juan Islands in two years. As they approached, the sounds of the blows reached us first, echoing over the glassy-flat waters. *Kawoof, kawoof, kawoof* – just the sound of them alone was enough to get the heart racing, but then the dorsal fins appeared. The first large group of maybe 25 whales was in party mode, breaching, swimming upside down, and spyhopping their way down the channel. Two whale-watching boats sat parked with their engines off several hundred yards away from the whales, and it was quiet enough to hear the people onboard cheering when K21 Cappuccino did a series of full-body breaches.

Next came a mixed group of J-Pod whales, pausing occasionally to opportunistically pursue a salmon on their way south. Since the deaths of matriarchs J2 Granny, J8 Spieden, and J14 Samish, the pod has been less inclined to hang out in their distinct matrilines, and instead seem to be segregating by age-sex class. It's bittersweet to see J31 Tsuchi, J40 Suttles, and J46 Star traveling together, as they are all females who lost their mothers at a young age. They are yet to have their own first successful calf, but carry much of the hope for the future of J-Pod, and are perhaps building the

key female-female relationships that will help them successfully raise young in the absence of their own mothers when
the time comes. Similarly, many of the young males have
banded together: J38 Cookie, J47 Notch, J49 T'ilem I'nges,
and L87 Onyx. While these J-Pod males still have living
mothers, they've formed a new social network for adopted
J-Pod member Onyx, who continues to show amazing resil-
ience after the death of not only his mother but numerous
surrogate mother figures.

The final group to pass is much closer to shore where we
are standing. It's now getting too dark to take photos, but
we can still clearly make out the black silhouettes of each
whale as they surface. There's K27 Deadhead, a mother
who has had one successful calf and at least one failed pregnancy, and was just recently announced as pregnant again
by the photogrammetry team who can monitor body condition from the air. She is just one of several whales pregnant at this time, but the low number of pregnancies carried to full term in the last ten years keeps us from getting
too excited about this news just yet. Following Deadhead
is her brother K25 Scoter, the latest whale to be looking
noticeably thin. In the words of John Durban, one member of the photogrammetry team, our increased knowledge
of these whales means we can now understand the "social
basis to vulnerability" of a given animal. It's thus not surprising to see Scoter suffering, as not only has he lost his
mother, but his sisters both have offspring and Deadhead
is further nutritionally taxed by her pregnancy. This means
fewer whales to share food with him, putting him at a
greater risk in these lean times.

One of the last pairs of whales to pass is J17 Princess Angeline and her youngest daughter J53 Kiki, the only remaining female calf from the recent baby boom. Somehow, she has beaten the odds, and while some of her cohorts didn't make it, she has thrived, is growing, and always seems to be exuberant. She, too, is a very key whale for the future of this population. As darkness falls over the San Juan Islands, we again lose sight of the whales, left only to hear their echoing breaths: *kawoof, kawoof, kawoof.*

The encounter was a simple one, like so many others. A few whale lovers on shore and on boats observing and photographing the Southern Residents, the most-watched group of killer whales in the world. The sighting epitomized so much of what is going on with this community: the excitement of seeing a youngster like Kiki doing well and a pregnant female in Deadhead coupled with the sadness of having gone another three years with no new births and seeing Scoter looking thin. This encounter marked the end of my eighteenth year watching these whales. I continue to hope that generations of orcas and humans alike will come together for these special encounters in the Salish Sea.

ACKNOWLEDGMENTS

THE INITIAL VISION FOR THIS book stemmed from a writing project with Shann Weston, Kari Koski, and Katie Jones. Their ideas and feedback during our meetings greatly shaped the project and got it off the ground.

From the first acoustics meeting where he recommended I apply (at age fifteen) for an internship at The Whale Museum, Rich Osborne has been a mentor and a friend. His early support for this book gave me the confidence to pursue it. As someone who has been around the Southern Residents from the early research days, his insight has been invaluable. Thank you for all the meetings, interviews, and emails that helped me tell this story.

It would have been impossible to write this book without the knowledge and stories so many people graciously shared with me through interviews, phone calls, lectures, and emails. For the time they gave me as well as their dedication to the whales, I thank Andrew Lees, Barbara Howitt, Binney Haenel, Bob Otis, Bonnie Gretz, Brad Hanson, Brett Soberg, Brian Goodremont, Bruce Stedman, Cindy Hansen, Dave Ellifrit, David Jamison, Howard Garrett, Howie Rosenfeld, James Taylor, Jenny Atkinson, Jessica Lundin, Jim Boran,

John Durban, John Ford, Katherine Ayres, Ken Balcomb, Lance Barrett-Lennard, Lynne Barre, Mark Anderson, Mike Ford, Noreen Ignelzi, Peter and Nancy Hardy, Ralph Munro, Richard Daly, Sara Heimlich, Sharon Grace, Stephen Raverty, Susan Berta, and Susan Vernon.

When my belief in this book wavered, I found support from several friends, including Brittany Bowles, Chris Verlinden, Heitor Crespo, Julie Woodruff, and Sara Hysong-Shimazu. You all inspired me to stick with it and convinced me I was the right person to tell this story. Sara also generously donated her talents to create the maps in this book.

The path to publication took an unexpected turn late in the game, but my vision for the book was still obtainable thanks to the last-minute support from Ana-Maria Brujban, Anonymous (x3), Alexandra Thomas, Alison Engle, Alissa Elderkin, Ariel Yseth, Binney Haenel, Bryan Dawley, Candance Calloway-Whiting, Chris Losee, Cindy Hansen, Connie Bickerton, David Neiwert, Donna and Bill Radcliffe, Emily Daggett, Emma Luck, Ginny Gensler, James Taylor, Jill Bliss, Julie Woodruff, Katie McLaughlin, Marlene Martinez, Michelle Borsz, Rainer and Vera Wieland, Sarabeth Bjorndahl, Shann Weston, Susan Andersson, Valerie Jones, and Whitney Neugebauer.

I appreciate the time and effort of numerous people, including several listed above, who read early versions of the manuscript before it took its final form. In particular, I want to express gratitude to my parents, Rainer and Vera Wieland, who have supported my writing career since I wrote my first orca story at age seven. I owe huge thanks to Regan Huff whose sharp editorial skills strengthened this manuscript more than I can say. The expertise of editors

Rebecca Brinbury and Amy Smith Bell helped further refine the text, as did the insightful comments from three anonymous reviewers.

This writing project spanned ten years. At just the right moment, my husband Jason Shields provided the encouragement and support necessary for me to focus and reach the finish line. Thank you, Jason, for your partnership in this and all things.

APPENDIX 1

Southern Resident Killer Whales, 1976–2018

THE FOLLOWING IS A LIST of all Southern Resident killer whales given an alphanumeric designation (ID) between 1976 and 2018. Common names, genders, presumed mothers, inferred fathers, and birth and death years are also listed. Orca Survey, now a program of the Center for Whale Research, has conducted an annual census of the Southern Resident killer whale population since 1976. Common names are from The Whale Museum's Orca Adoption Program. Possible mothers (indicated with a question mark) are based on observed associations by the Center for Whale Research. All fathers are inferred from genetic analyses (data from Mike Ford et al., 2018, "Inbreeding in an Endangered Killer Whale Population"). Birth years in parentheses are estimates from the Center for Whale Research. Where gender is unknown, a "U" is indicated.

ID	Name	Gender	Mother	Father	Birth Year	Death Year
J1	Ruffles	M			(1951)	2010
J2	Granny	F			(1911)	2016
J3	Merlin	M	J7?		(1953)	1995
J4	Mama	F	J8		(1957)	1995
J5	Saratoga	F	J9?		(1939)	1997
J6	Ralph	M	J8?		(1956)	1998
J7	Sucia	F			(1939)	1983
J8	Spieden	F			(1917)	2013
J9	Neah	F			(1917)	1985
J10	Tahoma	F	J9?		(1962)	1999
J11	Blossom	F	J4		(1972)	2008
J12	Sissy	F	J2?		(1935)	1996
J13	Unnamed	F	J5?		(1971)	1980
J14	Samish	F	J12	J1	1974	2016
J15	Unnamed	M	J4		1976	1981
J16	Slick	F	J7		(1972)	
J17	Princess Angeline	F	J5		1977	
J18	Everett	M	J10		1977	2000
J19	Shachi	F	J4		1979	
J20	Ewok	F	J10		1981	1998
J21	E.T.	U	J4		1982	1983
J22	Oreo	F	J10	J1	1985	
J23	Unnamed	M	J14		1987	1991
J24	Unnamed	U	J12		1972	1972
J25	Unnamed	U	J11		1988	1988
J26	Mike	M	J16		1991	
J27	Blackberry	M	J11	J1	1991	
J28	Polaris	F	J17	J1	1993	2016
J29	Unnamed	M	J19		1993	1993
J30	Riptide	M	J14		1995	2011
J31	Tsuchi	F	J11		1995	
J32	Rhapsody	F	J20	J1	1996	2014
J33	Keet	M	J16	J1	1996	2010
J34	Doublestuf	M	J22	L41	1998	2016
J35	Tahlequah	F	J17	L41	1998	
J36	Alki	F	J16	L41	1999	
J37	Hy'Shqa	F	J14	L41	2001	
J38	Cookie	M	J22		2003	
J39	Mako	M	J11	J1	2003	

ID	Name	Gender	Mother	Father	Birth Year	Death Year
J40	Suttles	F	J14	L41	2004	
J41	Eclipse	F	J19	J1	2005	
J42	Echo	F	J16	J26	2007	
J43	Unnamed	U	J14		2007	2007
J44	Moby	M	J17	L41	2009	
J45	Se-Yi-Chn	M	J14	L41	2009	
J46	Star	F	J28	J1	2009	
J47	Notch	M	J35		2010	
J48	Unnamed	U	J16		2011	2011
J49	T'ilem I'nges	M	J37	L79	2012	
J50	Scarlet	F	J16	L41	2014	2018
J51	Nova	M	J41		2015	
J52	Sonic	M	J36		2015	2017
J53	Kiki	F	J17	L41	2015	
J54	Dipper	M	J28		2015	2016
J55	Unnamed	U	J37?		2016	2016
K1	Taku	M	K7?		(1955)	1997
K2	Unnamed	M	K7?		(1953)	1974
K3	Sounder	F	K8?		(1957)	1998
K4	Morgan	F			(1933)	1999
K5	Sealth	M	K8?		(1953)	1991
K7	Lummi	F			(1910)	2008
K8	Tumwater	F			(1930)	1989
K11	Georgia	F			(1933)	2010
K12	Sequim	F	K4?		(1972)	
K13	Skagit	F	K11		(1972)	
K14	Lea	F	K3	J1	1977	
K15	Unnamed	U	K3		(1971)	1975
K16	Opus	F	K3		1985	
K17	Pacheena	M	K18?		(1966)	1994
K18	Kiska	F			(1948)	2003
K19	Neptune	M	K30?		(1953)	1984
K20	Spock	F	K13		1986	
K21	Cappuccino	M	K18		1986	
K22	Sekiu	F	K12		1987	
K23	Unnamed	U	K14		1988	1988
K24	Unnamed	U	K14		1990	1990
K25	Scoter	M	K13		1991	
K26	Lobo	M	K14		1993	

ID	Name	Gender	Mother	Father	Birth Year	Death Year
K27	Deadhead	F	K13	J1	1994	
K28	Raven	F	K12		1994	2006
K29	Sigurd	M	K3		1996	1998
K30	Unnamed	F			(1929)	1982
K31	Tatoosh	M	K12		1999	2005
K32	Unnamed	U	K16		2000	2000
K33	Tika	M	K22	L41	2001	
K34	Cali	M	K13	L41	2001	
K35	Sonata	M	K16	L41	2002	
K36	Yoda	F	K14	L41	2003	
K37	Rainshadow	M	K12	J1	2003	
K38	Comet	M	K20	L57	2004	
K39	Unnamed	U	K28		2006	2006
K40	Raggedy	F	K18?		(1963)	2012
K41	Unnamed	U	K22		2006	2006
K42	Kelp	M	K14	L41	2008	
K43	Saturna	F	K12	J1	2010	
K44	Ripple	M	K27	L78	2011	
K46	Unnamed	U	K18		1974	1981
L1	Oskar	M	L35		(1959)	2000
L2	Grace	F			(1960)	2012
L3	Oriana	F	L9?		(1948)	2002
L4	Sonar	F			(1949)	1996
L5	Tanya	F	L9?		(1964)	2012
L6	Podner	M	L2		(1962)	1983
L7	Canuck	F	L37?		(1961)	2010
L8	Moclips	M	L66?		(1958)	1977
L9	Hopi	F			(1931)	1996
L10	Okum	M	L12?		(1959)	1997
L11	Squirty	F	L12?		(1957)	2000
L12	Alexis	F			(1933)	2012
L13	Orpheus	M	L15?		(1952)	1980
L14	Cordy	M	L23?		(1972)	1989
L15	Gracie	F			(1930)	1981
L16	Unnamed	M	L7?		(1949)	1978
L20	Trident	M	L15?		(1955)	1982
L21	Ankh	F			(1950)	2008
L22	Spirit	F	L32		(1971)	
L23	Tsunami	F	L25?		(1952)	1982

ID	Name	Gender	Mother	Father	Birth Year	Death Year
L25	Ocean Sun	F			(1928)	
L26	Baba	F			(1956)	2013
L27	Ophelia	F	L4?		(1965)	2015
L28	Misky	F			(1949)	1994
L32	Olympia	F			(1955)	2005
L33	Chinook	M	L3?		(1963)	1995
L35	Victoria	F			(1942)	1996
L36	Unnamed	U	L45		1975	1975
L37	Kimo	F			(1933)	1984
L38	Dylan	M	L28?		(1965)	1998
L39	Orcan	M	L2		1975	2000
L41	Mega	M	L11		1978	
L42	Mozart	M	L11?		(1973)	1994
L43	Jellyroll	F	L37		(1972)	2006
L44	Leo	M	L32		1974	1998
L45	Asterix	F	L66		(1938)	1995
L47	Marina	F	L21		1974	
L48	Flash	U	L21		1977	1983
L49	Unnamed	U	L23?		1979	1980
L50	Shala	M	L35?		(1973)	1989
L51	Nootka	F	L3		(1973)	1999
L52	Salish	U	L26		1980	1983
L53	Lulu	F	L7		1977	2014
L54	Ino	F	L35		1977	
L55	Nugget	F	L4		1977	
L56	Disney	U	L32		1978	1981
L57	Faith	M	L45		1977	2008
L58	Sparky	M	L5		1980	2002
L59	Fred	U	L3		1979	1979
L60	Rascal	F	L26		(1972)	2002
L61	Astral	M	L4?		1973	1996
L62	Cetus	M	L27		1980	2000
L63	Scotia	M	L32		1984	1995
L64	Radar	U	L11		1985	1985
L65	Aquarius	F	L35		1984	1994
L66	Mata Hari	F			(1924)	1986
L67	Splash	F	L2		1985	2008
L68	Elwha	M	L27		1985	1995
L69	Sumner	U	L28		1984	1985

ID	Name	Gender	Mother	Father	Birth Year	Death Year
L71	Hugo	M	L26		1986	2006
L72	Racer	F	L43		1986	
L73	Flash	M	L5		1986	2010
L74	Saanich	M	L3		1986	2009
L75	Panda	F	L22		1986	1993
L76	Mowgli	U	L7		1987	1987
L77	Matia	F	L11		1987	
L78	Gaia	M	L2		1989	2012
L79	Skana	M	L22		1989	2013
L80	Odessa	U	L27		1990	1993
L81	Raina	M	L60		1990	1997
L82	Kasatka	F	L55		1990	
L83	Moonlight	F	L47		1990	
L84	Nyssa	M	L51		1990	
L85	Mystery	M	L28		1991	
L86	Surprise!	F	L4		1991	
L87	Onyx	M	L32		1992	
L88	Wavewalker	M	L2		1993	
L89	Solstice	M	L22	L41	1993	
L90	Ballena	F	L26		1993	
L91	Muncher	F	L47		1995	
L92	Crewser	M	L60		1995	
L93	Nerka	F	L27		1995	1998
L94	Calypso	F	L11		1995	
L95	Nigel	M	L43	L41	1996	2018
L96	Bernardo	M	L55		1996	1997
L97	Tweak	U	L51		1999	1999
L98	Luna	M	L67		1999	2006
L99	Unnamed	U	L47		2000	2000
L100	Indigo	M	L54	L41	2001	2014
L101	Aurora	U	L67	L41	2002	2008
L102	Unnamed	U	L47		2002	2002
L103	Lapis	F	L55	J1	2003	
L104	Domino	M	L43		2004	2006
L105	Fluke	M	L72	L57	2004	
L106	Pooka	M	L27	L41	2005	
L107	Unnamed	U	L47		2005	2005
L108	Coho	M	L54	J1	2006	
L109	Takoda	M	L55	J1	2007	

ID	Name	Gender	Mother	Father	Birth Year	Death Year
L110	Midnight	M	L83		2007	
L111	Unnamed	U	L47		2008	2008
L112	Sooke	F	L86	L41	2009	2012
L113	Cousteau	F	L94		2009	
L114	Unnamed	U	L77		2010	2010
L115	Mystic	M	L47		2010	
L116	Finn	M	L82	L41	2010	
L117	Keta	U	L54		2010	
L118	Jade	F	L55	J26	2011	
L119	Joy	F	L77	L78	2012	
L120	Unnamed	U	L86		2014	2014
L121	Windsong	M	L94		2015	
L122	Magic	M	L91		2015	
L123	Lazuli	M	L103		2015	

APPENDIX 2

Stranded Southern Resident Killer Whales

OCCASIONALLY, KILLER WHALES WASH UP deceased on the shoreline. The following are the known Southern Residents who have stranded, determined either by photo ID techniques or genetic testing.

Whale	Age	Date Discovered	Location	Cause of Death/Notes
L8 Moclips	19	August 5, 1977	Victoria, BC	Skeleton on display at The Whale Museum
Unk	Neonate	March 26, 1978		
Unk	Neonate	November 5, 1978		
L66 Mata Hari	62	August 14, 1986	Port Alberni, BC	
Unk	Neonate	October 7, 1986		
L14 Cordy	17	April 22, 1989	West coast of Vancouver Island	
J4 Mama	38	January 1, 1995	Vancouver Island	
L51 Nootka	26	September 25, 1999	Bentinck Island, BC	Prolapsed uterus
J18 Everett	23	March 18, 2000	Boundary Bay, BC	Bacterial infection (*Edwardsiella tarda*)
L60 Rascal	30	April 15, 2002	Long Beach, Washington	
L98 Luna	7	March 10, 2006	Nootka Sound, BC	Vessel strike, propeller from the tugboat *General Jackson*
Unk (J-Pod)	Neonate	July 26, 2008	Henry Island, Washington	Likely an aborted fetus. From genetic analysis the mother was likely J11 and the father was likely L57.

Whale	Age	Date Discovered	Location	Cause of Death/Notes
L112 Sooke	3	February 11, 2012	Long Beach, Washington	Massive blunt force trauma to the head and neck from an unknown source; no official cause of death given
Unk male	Neonate	January 2013	Dungeness Spit, Washington	Born alive but never seen alive; From genetic analysis the mother was likely J28 and the father was likely L41.
J32 Rhapsody	18	December 4, 2014	Comox, BC	Death of full-term fetus she was carrying caused a fatal infection (endometritis and maternal septicemia). From genetic analysis the father of the female fetus was likely L85.
Unk female	Neonate	March 23, 2016	Sooke, BC	Female calf under two weeks old; undocumented alive. From genetic analysis the father was likely K26.
L95 Nigel	20	March 31, 2016	Esperanza, BC	Fungal infection that entered his body through punctures from the recently applied satellite tag
J34 Doublestuf	18	December 20, 2016	Sechelt, BC	Blunt force trauma to the head; no official cause of death given

Selected Bibliography and Further Reading

Anchorage Daily News. "Killer Whales Corralled." March 8, 1976.

Ayres, Katherine L., Rebecca K. Booth, Jennifer A. Hempelmann, Kari L. Koski, Candice K. Emmons, Robin W. Baird, Kelley Balcomb-Bartok, M. Bradley Hanson, Michael J. Ford, and Samuel K. Wasser. "Distinguishing the Impacts of Inadequate Prey and Vessel Traffic on an Endangered Killer Whale (*Orcinus orca*) Population." *PLoS ONE* 7, no. 6 (2012). https://doi.org/10.1371/journal.pone.0036842.

Baird, Robin W. *Killer Whales of the World: Natural History and Conservation.* Vancouver, BC: Voyageur Press, 2002.

———. "Predators, Prey, and Play: Killer Whales and Other Marine Mammals." *Whalewatcher: Journal of the American Cetacean Society* 40, no. 1 (2011).

Baird, Robin W., and Pam J. Stacey. "Variation in Saddle Patch Pigmentation in Populations of Killer whales (*Orcinus orca*) from British Columbia, Alaska, and Washington State." *Canadian Journal of Zoology* 66, no. 11 (1988). https://doi.org/10.1139/z88-380.

Barre, Lynne. "How Do You Keep Killer Whales Away from an Oil Spill?" NOAA Office of Response and Restoration, February 8, 2016. http://response.restoration.noaa.gov/about/media/how-do-you-keep-killer-whales-away-oil-spill.html.

Barrett-Lennard, Lance. "Population Structure and Mating Patterns of Killer Whales (*Orcinus orca*) As Revealed by DNA Analysis." PhD diss., University of British Columbia, 2000.

Benedict, Audrey D., and Joseph K. Gaydos. *The Salish Sea: Jewel of the Pacific Northwest.* Seattle: Sasquatch Books, 2015.

Bigg, Michael A., Graeme M. Ellis, John K. Ford, and Kenneth C. Balcomb. *Killer Whales: A Study of Their Identification, Genealogy and Natural History in British Columbia and Washington State.* Nanaimo, BC: Phantom Press, 1987.

Bigg, Michael A., Peter Olesiuk, Graeme M. Ellis, John K. Ford, and Kenneth C. Balcomb. "Social Organization and Genealogy of Resident Killer Whales (*Orcinus Orca*) in the Coastal Waters of British Columbia and Washington State." Reports of the International Whaling Commission, Special Issue 12 (1990): 383–405.

Blackfish. Directed by Gabriela Cowperthwaite. Magnolia Pictures, 2013.

Brent, Lauren J. N., Daniel W. Franks, Emma A. Foster, Kenneth C. Balcomb, Michael A. Cant, and Darren P. Croft. "Ecological Knowledge, Leadership, and the Evolution of Menopause in Killer Whales." *Current Biology* 25 (2015): 1–5. http://dx.doi.org/10.1016/j.cub.2015.01.037.

Chasco, Brandon, Isaac C. Kaplan, Austen Thomas, Alejandro Acevedo-Gutierrez, Dawn Noren, Michael J. Ford, M. Bradley Hanson et al. "Competing Tradeoffs between Increasing Marine Mammal Predation and Fisheries Harvest of Chinook Salmon." *Scientific Reports* 7, no. 15439 (2017). http://doi.org/10.1038/s41598-017-14984-8.

———. "Estimates of Chinook Salmon Consumption in Washington State Inland Waters by Four Marine Mammal Predators from 1970 to 2015." *Canadian Journal of Fisheries and Aquatic Sciences* 74, no. 8 (2017): 1173–94. https://doi.org/10.1139/cjfas-2016-0203.

Croft, Darren P., Rufus A. Johnstone, Samuel Ellis, Stuart Nattrass, Daniel W. Franks, Lauren J. N. Brent, Sonia Mazzi, Kenneth C.

Balcomb, John K. B. Ford, and Michael A. Cant. "Reproductive Conflict and the Evolution of Menopause in Killer Whales." *Current Biology* 27 (2017): 1–7. http://dx.doi.org/10.1016/j.cub.2016.12.015

Deecke, Volker B., John K. B. Ford, and Paul Spong. "Dialect Change in Resident Killer Whales: Implications for Vocal Learning and Cultural Transmission." *Animal Behaviour* 60, no. 5 (2000): 629–38. http://doi.org/10.1006/anbe.2000.1454.

Durban, John W., and Kim M. Parsons. "Laser-metrics of Free-ranging Killer Whales." *Marine Mammal Science* 22, no. 3 (2006): 735–43. https://doi.org/10.1111/j.1748-7692.2006.00068.x.

Durban, John, Holly Fearnbach, Dave Ellifrit, and Ken Balcomb. "Size and Body Condition of Southern Resident Killer Whales." Contract report to the Northwest Regional Office, National Marine Fisheries Service. 2009.

Erbe, Christine. "Underwater Noise of Whale-Watching Boats and Potential Effects on Killer Whales (*Orcinus orca*), Based on an Acoustic Impact Model." *Marine Mammal Science* 18, no. 2 (2002): 394–418. https://doi.org/10.1111/j.1748-7692.2002.tb01045.x.

Fearnbach, Holly, John W. Durban, Dave K. Ellifrit, and Ken C. Balcomb. "Size and Long-Term Growth Trends of Endangered Fish-Eating Killer Whales." *Endangered Species Research* 13 (2011): 173–80. https://doi.org/10.3354/esr00330.

Fisheries and Oceans Canada. *Recovery Strategy for Northern and Southern Resident Killer Whales* (Orcinus orca) *in Canada*. Ottawa: Fisheries and Oceans Canada, *Species at Risk Act* Recovery Strategy Series, 2011. ix + 80 pp.

Foote, Andrew D., Richard W. Osborne, and A. Rus Hoelzel. "Whale-call Response to Masking Boat Noise." *Nature* 428 (2004): 910. https://doi.org/10.1038/428910a.

Ford, John K. B. "Acoustic Behavior of Resident Killer Whales (*Orcinus orca*) off Vancouver Island, British Columbia." *Canadian Journal of Zoology* 67, no. 3 (1989): 727–45. https://doi.org/10.1139/z89-105.

————. "Vocal Traditions among Resident Killer Whales (*Orcinus orca*) in Coastal Waters of British Columbia." *Canadian Journal of Zoology* 69, no. 6 (1991): 1454–83. https://doi.org/10.1139/z91-206.

Ford, John K. B., and Graeme M. Ellis. "Selective Foraging by Fish-Eating Killer Whales *Orcinus orca* in British Columbia." *Marine Ecology Progress Series* 316 (2006): 185–99. https://doi.org/10.3354/meps316185.

————. *Transients*. Vancouver: University of British Columbia Press, 1999.

Ford, John K. B., Graeme M. Ellis, Lance G. Barrett-Lennard, Alexandra B. Morton, Rod S. Palm, and Kenneth C. Balcomb. "Dietary Specialization in Two Sympatric Populations of Killer Whales (*Orcinus orca*) in Coastal British Columbia and Adjacent Waters." *Canadian Journal of Zoology* 76 (1998): 1456–71. https://doi.org/10.1139/z98-089.

Ford, John K. B., Graeme M. Ellis, Peter F. Olesiuk, and Kenneth C. Balcomb. "Linking Killer Whale Survival and Prey Abundance: Food Limitation in the Oceans' Apex Predator?" *Biology Letters* 6 (2010): 139–42. https://doi.org/10.1098/rsbl.2009.0468.

Ford, Michael J., Jennifer Hempelmann, M. Bradley Hanson, Katherine L. Ayres, Robin W. Baird, Candice K. Emmons, Jessica I. Lundin et al. "Estimating of a Killer Whale (*Orcinus orca*) Populations' Diet Using Sequencing Analysis of DNA from Feces." *PLoS ONE* 11, no. 1 (2016). https://doi.org/10.1371/journal.pone.0144956.

Ford, Michael J., K. M. Parsons, E. J. Ward, J. A. Hempelmann, C. K. Emmons, M. Bradley Hanson, K. C. Balcomb, and L. K. Park. "Inbreeding in an Endangered Killer Whale Population." *Animal Conservation* (2018). https://doi.org/10.1111/acv.12413.

Ford, Michael J., M. Bradley Hanson, Jennifer A. Hempelmann, Katherine L. Ayres, Candice K. Emmons, Gregory S. Schorr, Robin W. Baird et al. "Inferred Paternity and Male Reproductive Success in a Killer Whale (*Orcinus orca*) Population." *Journal of Heredity* 102, no. 5 (2011): 537–53. https://doi.org/10.1093/jhered/esr067.

Foster, Emma A., Daniel W. Franks, Sonia Mazzi, Safi K. Darden, Ken C. Balcomb, John K. B. Ford, and Darren P. Croft. "Adaptive Prolonged Postreproductive Life Span in Killer Whales." *Science* 337, no. 6100 (2012): 1313. http://doi.org/126/science.1224198.

Francis, Daniel, and Gil Hewlett. *Operation Orca: Springer, Luna, and the Struggle to Save West Coast Killer Whales*. Madeira Park, BC: Harbour Publishing, 2007.

Gaydos, Joseph K., and Scott F. Pearson. "Birds and Mammals That Depend on the Salish Sea: A Compilation." *Northwestern Naturalist* 92 (2011): 79–94.

Griffin, Ted. *Namu: Quest for the Killer Whale*. Seattle: Gryphon West Publications, 1982.

Hanson, M. Bradley, Robin W. Baird, John K. B. Ford, Jennifer Hempelmann-Halos, Donald M. Van Doornik, John R. Candy, Candice K. Emmons et al. "Species and Stock Identification of Prey Consumed by Endangered Southern Resident Killer Whales in Their Summer Range." *Endangered Species Research* 11 (2010): 69–82. https://doi.org/10.3354/esr00263.

Hargrove, John. *Beneath the Surface: Killer Whales, SeaWorld, and the Truth beyond Blackfish*. New York: St. Martin's Griffin, 2015.

Hawley, Steven. *Recovering a Lost River: Removing Dams, Rewilding Salmon, Revitalizing Communities*. Boston, MA: Beacon Press, 2011.

Heimlich-Boran, Sara. "Association Patterns and Social Dynamics of Killer Whales (*Orcinus orca*) in Greater Puget Sound." MA thesis, San Jose State University, 1988. http://aquaticcommons.org/id/eprint/359.

Hilborn, Ray, Sean P. Cox, Francis M. D. Gulland, David G. Hankin, Tom Hobbs, Daniel E. Schindler, and Andrew W. Trites. 2012. "The Effects of Salmon Fisheries on Southern Resident Killer Whales: Final Report of the Independent Science Panel." Prepared with the assistance of D. R. Marmorek and A. W. Hall, ESSA Technologies Ltd., Vancouver BC, for National Marine Fisheries Service (Seattle, WA) and Fisheries and Oceans Canada (Vancouver, BC). xv + 61 pp + appendixes.

Holt, Marla M., Dawn P. Noren, and Candice K. Emmons. "Effects of Noise Levels and Call Types on the Source Levels of Killer Whale Calls." *Journal of the Acoustic Society of America* 130, no. 5 (2011): 3100–06. http://dx.doi.org/10.1121/1.3641446.

Holt, Marla M., Dawn P. Noren, Val Veirs, Candice K. Emmons, and Scott Veirs. "Speaking Up: Killer Whales (*Orcinus orca*) Increase Their Call Amplitude in Response to Vessel Noise." *Journal of the Acoustic Society of America* 125, no. 1 (2009): 27–32. http://asa.scitation.org/doi/10.1121/1.3040028.

Houghton, Juliana, Marla M. Holt, Deborah A. Giles, M. Bradley Hanson, Candice K. Emmons, Jeffrey T. Hogan, Trevor A. Branch, and Glenn R. VanBlaricom. "The Relationship between Vessel Traffic and Noise Levels Received by Killer Whales (*Orcinus orca*)." *PLoS ONE* 10, no. 12 (2015). https://doi.org/10.1371/journal.pone.0140119.

Hoyt, Erich. *Orca: The Whale Called Killer.* Revised edition. Richmond Hill, Ontario: Firefly Books, 1990.

Kirby, David. *Death and SeaWorld: Shamu and the Dark Side of Killer Whales in Captivity.* New York: St. Martin's Press, 2012.

Kirkevold, Barbara C., and Joan S. Lockard, eds. *Behavioral Biology of Killer Whales.* New York: Alan R. Liss, Inc., 1986.

Kitsap Sun. "Dyes Inlet Orcas—Ten Years Later," 2007. http://data.kitsapsun.com/projects/story/dyes-inlet-orcas-ten-years-later/ (accessed August 29, 2012).

Koski, Kari L., and Rich W. Osborne. "The Evolution of Adaptive Management Practices for Vessel-based Wildlife Viewing in the Boundary Waters of British Columbia and Washington State: From Voluntary Guidelines to Regulations." Paper presented at the Puget Sound–Georgia Basin Research Conference, Seattle, March 2005.

Krahn Margaret M., M. Bradley Hanson, Gregory S. Schorr, Candice K. Emmons, Douglas G. Burrows, Jennie L. Bolton, Robin W. Baird, and Gina M. Ylitalo. "Effects of Age, Sex, and Reproductive Status on Persistent Organic Pollutant Concentrations in 'Southern Resident' Killer Whales." *Marine*

Pollution Bulletin 58, no. 10 (2009): 1522–29. https://doi.org/10.1016/j.marpolbul.2009.05.014.

Krahn, Margaret M., M. Bradley Hanson, Robin W. Baird, Richard H. Boyer, Douglas G. Burrows, Candice K. Emmons, John K. B. Ford et al. "Persistent Organic Pollutants and Stable Isotopes in Biopsy Samples (2004/2006) from Southern Resident Killer Whales." *Marine Pollution Bulletin* 51, no. 12 (2007): 1903–11. https://doi.org/10.1016/j.marpolbul.2007.08.015.

Lacy, Robert C., Rob Williams, Erin Ashe, Kenneth C. Balcomb, Lauren J. N. Brent, Christopher W. Clark, Darren P. Croft et al. "Evaluating Anthropogenic Threats to Endangered Killer Whales to Inform Effective Recovery Plans." *Scientific Reports* 7 (2017). https://doi.org/10.1038/s41598-017-14471-0.

Lackey, Robert T. "Restoring Wild Salmon to the Pacific Northwest: Chasing an Illusion?" In *What We Don't Know about Pacific Northwest Fish Runs: An Inquiry into Decision-Making*, edited by Patricia Koss and Mike Katz, 93–143. Portland, OR: Portland State University, 2000.

Lewiston Morning Tribune. "Capture of Killer Whales Defended by Federal Vet." March 30, 1976.

Lichatowich, Jim. *Salmon, People, Place.* Corvallis: Oregon State University Press, 2013.

Leiren-Young, Mark. *The Killer Whale Who Changed the World.* Vancouver, BC: Greystone Books, 2016.

Lolita: Slave to Entertainment. Directed by Tim Gorski. Rattle the Cage Productions, 2013.

MacDuffee, M., A. R. Rosenberger, R. Dixon, A. Jarvela Rosenberger, C. H. Fox, and P. C. Paquet. *Our Threatened Coast: Nature and Shared Benefits in the Salish Sea.* Sidney, BC: Raincoast Conservation Foundation, 2016.

Matkin, C. O., E. L. Saulitis, G. M. Ellis, P. Olesiuk, and S. D. Rice. "Ongoing Population-level Impacts on Killer Whales *Orcinus Orca* Following the *Exxon Valdez* Oil Spill in Prince William Sound, Alaska." *Marine Ecology Progress Series* 356 (2008): 269–81. https://doi.org/10.3354/meps07273.

McKervill, Hugh W. *The Salmon People*. Vancouver, BC: Whitecap Books, 2015.

Morin, Phillip A., Frederick I. Archer, Andrew D. Foote, Julia Vilstrup, Eric E. Allen, Paul Wade, John Durban et al. "Complete Mitochondrial Genome Phylogeographic Analysis of Killer Whales (*Orcinus orca*) Indicates Multiple Species." *Genome Research* 20 (2010): 908–16. www.genome.org/cgi/doi/10.1101/gr.102954.109.

Morton, Alexandra. *Listening to Whales: What the Orcas Have Taught Us*. New York: Ballantine Books, 2002.

National Marine Fisheries Service. *Recovery Plan for Southern Resident Killer Whales* (Orcinus orca). Seattle: National Marine Fisheries Service, Northwest Region, 2008.

———. *Southern Resident Killer Whales: Ten Years of Research and Conservation*. Special Report. Seattle: National Marine Fisheries Service, Northwest Region, 2014.

Olesiuk, Peter F., Mike A. Bigg, and Graeme M. Ellis. "Life History and Population Dynamics of Resident Killer Whales in the Coastal Waters of British Columbia and Washington State." *Report to the International Whaling Commission*. Special Issue 12 (1990).

Parfit, Michael, and Suzanne Chisholm. *The Lost Whale: The True Story of the Orca Named Luna*. New York: St. Martin's Press, 2013.

Pielou, Evelyn C. *After the Ice Age: The Return of Life to Glaciated North America*. Chicago: University of Chicago Press, 1991.

Pitman, Robert L., ed. "Killer Whale: The Top, Top Predator." *Whalewatcher: Journal of the American Cetacean Society* 40, no. 1 (2011).

Pollard, Sandra. *Puget Sound Whales for Sale*. Mount Pleasant, SC: Arcadia Publishing, 2014.

———. *A Puget Sound Orca in Captivity: The Fight to Bring Lolita Home*. Mount Pleasant, SC: Arcadia Publishing, 2019.

Rayne, Sierra M., Michael G. Ikonomou, Peter S. Ross, Graeme M. Ellis, and Lance G. Barrett-Lennard. "PBDEs, PBBs, and PCNs in Three Communities of Free-Ranging Killer Whales (*Orcinus orca*) from the Northeastern Pacific Ocean." *Environmental*

Science and Technology 38, no. 16 (2004): 4293–99. http://pubs. acs.org/doi/abs/10.1021/es049501.

Rehn, Nicola, Olga A. Filatova, John W. Durban, and Andrew D. Foote. "Cross-cultural and Cross-ecotype Production of a Killer Whale 'Excitement' Call Suggests Universality." *Naturwissenschaften* 98 (2011): 1–6. https://doi.org/10.1007/s00114-010-0732-5.

Reisch, Rüdiger, Lance G. Barrett-Lennard, Graeme M. Ellis, John K. B. Ford, and Volker B. Deecke. "Cultural Traditions and the Evolution of Reproductive Isolation: Ecological Speciation in Killer Whales?" *Biological Journal of the Linnean Society* 106 (2012): 1–17. https://doi.org/10.1111/j.1095-8312.2012.01872.x.

Rendell, Luke, and Hal Whitehead. "Culture in Whales and Dolphins." *Behavioral and Brain Sciences* 24 (2001): 309–82.

——. *The Cultural Lives of Whales and Dolphins.* Chicago: University of Chicago Press, 2014.

Rose, Naomi Anne. "The Social Dynamics of Male Killer Whales, *Orcinus orca*, in Johnstone Strait, British Columbia." PhD diss., University of California Santa Cruz, 1992.

Ross, Peter S. "Fireproof Killer Whales (*Orcinus orca*): Flame-Retardant Chemicals and the Conservation Imperative in the Charismatic Icon of British Columbia, Canada." *Canadian Journal of Fisheries and Aquatic Sciences* 63, no. 1 (2006): 224–34. https://doi.org/10.1139/f05-244.

Ross, Peter S., Graeme M. Ellis, Michael G. Ikonomou, Lance G. Barrett-Lennard, and R. F. Addison. "High PCB Concentrations in Free-Ranging Pacific Killer Whales, *Orcinus orca*: Effects of Age, Sex, and Dietary Preference." *Marine Pollution Bulletin* 40, no. 6 (2000): 504–15.

Safina, Carl. *Beyond Words: What Animals Think and Feel.* New York: Picador, 2015.

Saulitus, Eva. *Into Great Silence: A Memoir of Discovery and Loss among Vanishing Orcas.* Boston: Beacon Press, 2013.

Shields, Monika W., Jimmie Lindell, and Julie Woodruff. "Declining Spring Usage of Core Habitat by Endangered Fish-eating Killer Whales Reflects Decreased Availability of their Primary Prey."

Pacific Conservation Biology 24 (2018): 189–93. https://doi. org/10.1071/PC17041.

Spokesman Review. "Scientists Don't Agree on Killer Whale Studies." March 24, 1976.

Stacey, Pam J., and Robin W. Baird. "Birth of a 'Resident' Killer Whale off Victoria, British Columbia, Canada." *Marine Mammal Science* 13, no. 3 (1997): 504–8.

Stevens, Tracy A., Deborah D. Duffield, Edward D. Asper, K. Gilbey Hewlett, Al Bolz, Laurie J. Gage, and Gregory D. Bossart. "Preliminary Findings of Restriction Fragment Differences in Mitochondrial DNA among Killer Whales (*Orcinus orca*)." *Canadian Journal of Zoology* 67, no. 10 (1989): 2592–95. https:// doi.org/10.1139/z89-365.

Towers, Jared R., Muriel J. Hallé, Helena K. Symonds, Gary J. Sutton, Alexandra B. Morton, Paul Spong, James P. Borrowman, and John K. B. Ford. "Infanticide in a Mammal-eating Killer Whale Population." *Scientific Reports* 8 (2018). https://doi.org/10.1038/ s41598-018-22714-x.

Veirs, Scott, Val Veirs, and Jason Wood. "Ship Noise Extends to Frequencies Used tor Echolocation by Endangered Killer Whales." *PeerJ* 4:e1657 (2016). https://doi.org/10.7717/peerj.1657.

Vélez-Espino, L. Antonio, John K. B. Ford, H. Andres Araujo, Graeme Ellis, Charles K. Parken, and Rishi Sharma. "Relative Importance of Chinook Salmon Abundance on Resident Killer Whale Population Growth and Viability." Aquatic Conservation: Marine and Freshwater Ecosystems 25, no. 6 (2014): 756–80. https://doi.org/10.1002/ aqc.2494.

Victoria Advocate. "Judge Sets 'Killer' Whales Free." March 13, 1976.

Ward, Eric J., Kim Parsons, Elizabeth E. Holmes, Ken C. Balcomb, and John K. B. Ford. "The Role of Menopause and Reproductive Senescence in a Long-Lived Social Mammal." *Frontiers in Zoology* 6, no. 4 (2009). https://doi.org/10.1186/1742-9994-6-4.

Wasser, Samuel W., Jessica I. Lundin, Katherine Ayres, Elizabeth Seely, Deborah Giles, Kenneth Balcomb, Jennifer Hempelmann, Kim Parsons, and Rebecca Booth. "Population Growth Is

Limited by Nutritional Impacts on Pregnancy Success in Endangered Southern Resident Killer Whales (*Orcinus orca*). *PLOS ONE* 12, no. 6 (2017): e0179824. https://doi.org/10.1371/journal.pone.0179824.

Webber, Burt. "Naming the Salish Sea." Western Washington University, 2012. www.wwu.edu/salishsea/history.shtml (accessed March 5, 2015).

The Whale. Directed by Michael Parfit and Suzanne Chisholm. Mountainside Films, 2011.

Whitehead, Hal, Luke Rendell, Richard W. Osborne, and Bernd Würsig. "Culture and Conservation in Non-Humans with Reference to Whales and Dolphins: Review and New Directions." *Biological Conservation* 120, no. 3 (2004): 427–37. https://doi.org/10.1016/j.biocon.2004.03.017.

Williams, Rob, Andrew W. Trites, and David E. Bain. "Behavioural Responses of Killer Whales (*Orcinus orca*) to Whale-Watching Boats: Opportunistic Observations and Experimental Approaches." *Journal of the Zoological Society of London* 256, no. 2 (2002): 255–70. https://doi.org/10.1017/S0952836902000298.

Wright, Brianna M., Eva H. Stredulinsky, Graeme M. Ellis, and John K. B. Ford. "Kin-directed Food Sharing Promotes Lifetime Natal Philopatry of Both Sexes in a Population of Fish-Eating Killer Whales, *Orcinus orca*." *Animal Behavior* 115 (2016): 81–95. http://dx.doi.org/10.1016/j.anbehav.2016.02.025.

Yates, Steve. *Orcas, Eagles, and Kings: The Natural History of Puget Sound and Georgia Strait.* Seattle: Primavera Press, 1992.

INDEX

Note: page numbers in *italics* refer to illustrations and tables.

CPSIA information can be obtained
at www.ICGtesting.com
Printed in the USA
BVHW031742060319
541951BV00001B/9/P